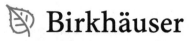

Progress in Mathematical Physics

Volume 71

More information about this series at http://www.springer.com/series/4813

Bertrand Duplantier • Vincent Rivasseau
Jean-Nöel Fuchs

Editors

Dirac Matter

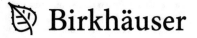 Birkhäuser

Editors
Bertrand Duplantier
Service de Physique Théorique
CEA Saclay
Gif Sur Yvette Cedex, France

Vincent Rivasseau
Laboratoire de Physique Théorique
Université Paris-Sud
Orsay, France

Jean-Nöel Fuchs
Laboratoire de Physique
Université Pierre-et-Marie Curie
Paris, France

ISSN 1544-9998 ISSN 2197-1846 (electronic)
Progress in Mathematical Physics
ISBN 978-3-319-81311-0 ISBN 978-3-319-32536-1 (eBook)
DOI 10.1007/978-3-319-32536-1

Mathematics Subject Classification (2010): 82-01, 82-02, 81V70, 35Q41

Printed on acid-free paper

This book is published under the trade name Birkhäuser, www.birkhauser-science.com
The registered company is Springer International Publishing AG
The registered company address is: Gewerbestrasse 11, 6330 Cham, Switzerland

Contents

Contributors

Philip Kim
Department of Physics, Harvard University, Cambridge, Massachussets, USA

Mark Goerbig and Gilles Montambaux
Laboratoire de Physique des Solides, Université Paris Sud, Orsay, France

Chuan Li, Sophie Guéron and Hélène Bouchiat
Laboratoire de Physique des Solides, Université Paris Sud, Orsay, France

Laurent Lévy
Institut Néel, Université Grenoble Alpes and CNRS, Grenoble, France

David Carpentier
Laboratoire de Physique, Ecole Normale Supérieure de Lyon, Lyon, France

Foreword

This book is the fifteenth in a series of Proceedings for the *Séminaire Poincaré*, which is directed towards a broad audience of physicists, mathematicians, and philosophers of science.

The goal of this Seminar is to provide up-to-date information about general topics of great interest in physics. Both the theoretical and experimental aspects of the topic are covered, generally with some historical background. Inspired by the *Nicolas Bourbaki Seminar* in mathematics, hence nicknamed *"Bourbaphy"*, the Poincaré Seminar is held twice a year at the Institut Henri Poincaré in Paris, with written contributions prepared in advance. Particular care is devoted to the pedagogical nature of the presentations, so that they may be accessible to a large audience of scientists.

This new volume of the Poincaré Seminar Series, *Dirac Matter*, corresponds to the eighteenth such seminar, held on June 28, 2014. Its aim is to provide a general introduction to the physics of materials, the conduction properties of which are due to electrons described by the Dirac equation. The latter was introduced by Paul Dirac in 1928 in order to wed quantum mechanics and special relativity so as to describe a fast moving electron beyond the non-relativistic Schrödinger equation. This outstanding contribution was recognized by the award of the 1933 Nobel Prize in Physics to Dirac. Although in high-energy physics Dirac's equation was long ago superseded by quantum field theory as the description of fundamental particles, it was recently resurrected as the correct low-energy description of many condensed matter systems, such as graphene and topological insulators. The present volume explains why Dirac still matters.

The first article, entitled "Graphene and Relativistic Quantum Physics", and written by the experimental pioneer, PHILIP KIM, is devoted to graphene, a two-dimensional form of crystalline carbon organized in a honeycomb lattice. The author starts by recounting the path that lead to the discovery in 2004-2005 by Kostya Novoselov and Andre Geim of this new type of crystals, whose thickness is that of a single atom. The discovery of graphene and its use as an electronic device was soon after recognized by the award to the pair of the 2010 Nobel Prize. Kim then goes on to explain the surprising behavior of electrons in this material by showing how the massless Dirac equation emerges as their low-energy effective description. Indeed, the band structure of graphene is such that the conduction and valence bands touch at two inequivalent contact points which form a degenerate Fermi surface. In the vicinity of these contact points, electrons are not described by a non-relativistic Schrödinger equation with an effective band mass – as is usually the case in metals or semiconductors – but by a relativistic-like Dirac equation for massless particles. In other words, electrons in graphene propagate with a dispersion relation, i.e., an energy as a function of lattice momentum, similar to that of photons and now known as a *Dirac cone*, albeit with a velocity which is not that of light in vacuum but three hundred times smaller and known as the Fermi velocity. Electrons in graphene are therefore not truly relativistic, and still have their usual gravitational mass, but their motion *appears* to be ultra-relativistic, i.e., with a vanishing inertial mass and with a corresponding characteristic Fermi velocity, the slope of the Dirac cones.

After setting the stage, Kim then shows how transport experiments are able to probe the peculiar properties of electrons in graphene. It turned out to be a spectacular playground to test the predictions of relativistic quantum mechanics. For example, the nearly century-old tunneling paradox, first formulated by Oskar Klein shortly after the discovery of the Dirac equation, could be probed in several transport experiments across heterojunctions in graphene. Also, a new version of the integer quantum Hall effect was measured in graphene and dubbed the relativistic quantum Hall effect. It features an unusual sequence of quantized plateaus in the transverse electric resistance when the graphene sheet is placed in a perpendicular magnetic field at low temperature. The peculiarity of the quantum Hall effect discovered in graphene by Kim and co-workers – and independently by Novoselov, Geim and co-workers – can be traced back to the existence of a so-called Berry phase of π. This geometric phase is acquired by the wave-function of an electron upon its circling around the Dirac cone and is related to the fact that the wave-function is not a scalar but a (pseudo-)spinor. In graphene, this pseudo-spin 1/2 has its origin in the structure of the honeycomb lattice which is made of *two* interpenetrating triangular lattices.

The second article offers a review entitled "Dirac Fermions in Condensed Matter and Beyond", written by two prominent theoreticians, MARK GOERBIG and GILLES MONTAMBAUX. Their aim is to show that Dirac type physics is not limited to graphene but exists in other materials, now collectively known as "Dirac matter". These materials can be organic salts under pressure, or transition metal dichalcogenides, or even surface states of three-dimensional topological insulators (see below), and are not limited to solid state samples. Dirac physics indeed also emerges in artificial systems such as cold atomic gases in honeycomb-like optical lattices, in microwave or photonic crystals, in polariton crystals, etc.

Goerbig and Montambaux then proceed to show that the peculiar topology of graphene's band structure can be revealed by deforming the honeycomb lattice, e.g., by applying uniaxial strain. Indeed, Dirac cones in the energy spectrum exist in pairs, usually known as valleys. These can also be seen as pairs of topological defects in the electronic wave-function. These defects are characterized by a topological charge, which is essentially their Berry phase. Upon stretching the microscopic lattice, it is possible to make these Dirac cones move and meet in reciprocal space . Upon meeting, the two defects annihilate (just as a vortex and anti-vortex do) and an energy gap opens in the band structure: Dirac cones disappear. Goerbig and Montambaux give a thorough description of this merging transition and discuss various experiments in which it was measured.

The third contribution, entitled "Quantum Transport in Graphene: Impurity Scattering as a probe of the Dirac Spectrum", is due to HÉLÈNE BOUCHIAT, a leading experimentalist in the field of mesoscopic physics, and her colleagues SOPHIE GUÉRON and CHUAN LI. They show how measuring electrical transport in real graphene devices – contaminated by impurities and hence exhibiting a diffusive regime – allows one to probe the Dirac nature of electrons. In particular, magneto-transport allows one to access two different scattering times of electrons on impurities. The ratio of these scattering times reveals that electrons in graphene do not back-scatter, which is the feature of the massless Dirac equation at the origin of Klein tunneling. Bouchiat, Guéron and Li have also performed experiments on graphene devices coupled to superconducting contacts. They have shown the possible occurrence of specular Andreev reflection, which is unique to undoped graphene and its Dirac cone, and replaces the more conventional Andreev retro-reflection well-known to occur at the interface between a superconducting and a normal metal. The authors end by envisioning future functionalities of graphene that could occur due to the deposition of adsorbates such as molecules in order to induce strong spin-orbit coupling in graphene. This could lead to the realization of a quantum spin Hall effect in graphene, as proposed by Charles Kane and Eugene Mele.

After these articles mainly concerned with the properties of graphene-like two-dimensional systems featuring Dirac electrons in their bulk, the last two contributions introduce topological insulators. These materials are band insulators that host conducting states at their surfaces or edges. In particular, three-dimensional topological insulators are bulk insulators with a two-dimensional Dirac metal at their surface.

In the authoritative "Experimental Signatures of Topological Insulators", LAURENT LÉVY reviews recent experimental progress in the fabrication and characterization of three-dimensional topological insulators. He focuses on mercury-telluride samples, which are crystals with a large spin-orbit coupling. In the absence of strain, this material is a semi-metal, that conducts electricity in its bulk. Under strain, its band structure changes into that of a bulk insulator with conducting surface states, i.e., a topological insulator. The surface states are probed by angle-resolved photo-emission spectroscopy (ARPES) and are shown to have a dispersion relation with the shape of a single Dirac cone similar to that of graphene. In addition, Lévy shows that when these strained samples are put into contact, the electrical current is carried by the surface states. The magneto-conductance features at least two signatures of Dirac-type physics: the unusual quantization of Landau levels seen in the Shubnikov-de Haas oscillations (at large magnetic field) and the weak anti-localization (at small magnetic field). These two effects are attributed to the Berry phase of π of the electrons at the surface of the sample. These experiments convincingly show that the surface of a three-dimensional topological insulator hosts a two-dimensional massless Dirac metal.

This volume ends with the contribution of DAVID CARPENTIER, entitled "Topology of Bands in Solids: From Insulators to Dirac Matter". This theoretical review of the field of topological insulators emphasizes the fundamental relation between these insulators and conductors described by a Dirac equation. The author starts by briefly surveying traditional band theory which is restricted to classifying energy spectra. He then proceeds by explaining how band theory hides a geometric description in its Bloch wave functions. This is explained in terms of Berry phases and parallel transport. The topological classification of the Bloch states in terms of invariants such as Chern numbers thus allows one to refine the classification of band insulators and to distinguish between trivial and topological insulators.

He also proves that gapless conducting edge/surface states have to exist at the boundary between a topological and a trivial insulator, as this border corresponds to a change of topology, which by nature cannot be continuous but requires the local closing of the energy gap. The critical states that exist at the edge/surface are then shown to be described by a massless Dirac equation. This illuminating review ends with perspectives on the study of three-dimensional semi-metals described by a Weyl equation, which is akin to the Dirac equation. These materials have point-like Fermi surfaces, like graphene but now in three space dimensions, which can be topologically classified as defects, as shown by Grigori Volovik.

This book, by the breadth of topics covered in Dirac's unforeseen legacy in condensed matter physics, should be of broad interest to physicists, mathematicians, and historians of science. We further hope that the continued publication of this series of Proceedings will serve the scientific community, at both the professional and graduate levels. We thank the COMMISSARIAT À L'ÉNERGIE ATOMIQUE ET AUX ÉNERGIES ALTERNATIVES (Division des Sciences de la Matière), the DANIEL IAGOLNITZER FOUNDATION, and the ÉCOLE POLYTECHNIQUE for sponsoring this Seminar. Special thanks are due to Chantal DELONGEAS for the preparation of the manuscript.

Saclay, Paris
& Orsay,
January 2016

BERTRAND DUPLANTIER
Institut de
Physique Théorique
Saclay, CEA
Université Paris-Saclay
Gif-sur-Yvette, France
bertrand.duplantier@cea.fr

JEAN-NÖEL FUCHS
Laboratoire de
Physique Théorique
de la Matière Condensée
Université Pierre-et-Marie Curie
Paris, France
fuchs@lptmc.jussieu.fr

VINCENT RIVASSEAU
Laboratoire de
Physique Théorique
Université Paris-Saclay
Orsay, France
rivass@th.u-psud.fr

Dirac Matter, 1–23
© 2016 Springer Basel AG

Graphene and Relativistic Quantum Physics

Philip Kim

Abstract. The honeycomb lattice structure of graphene requires an additional degree of freedom, termed as pseudo spin, to describe the orbital wave functions sitting in two different sublattices of the honeycomb lattice. In the low energy spectrum of graphene near the charge neutrality point, where the linear carrier dispersion mimics the quasi-relativistic dispersion relation, pseudo spin replaces the role of real spin in the usual Dirac Fermion spectrum. The exotic quantum transport behavior discovered in graphene, such as the unusual half-integer quantum Hall effect and Klein tunneling effect, are a direct consequence of the pseudo spin rotation. In this chapter we will discuss the non-trivial Berry phase arising from the pseudo spin rotation in monolayer graphene under a magnetic field and its experimental consequences.

1. Introduction

Graphene is one atom thick layer of carbon atoms arranged in a honeycomb lattice. This unique two-dimensional (2D) crystal structures provide an additional degree of freedom, termed as pseudo spin, to describe the orbital wave functions sitting in two different sublattices of the honeycomb lattice. In the low energy spectrum of graphene near the charge neutrality point, where the linear carrier dispersion mimics the "quasi-relativistic" dispersion relation, pseudo spin replaces the role of real spin in the usual relativistic Fermion energy spectrum. The exotic quantum transport behavior discovered in graphene, such as the unusual half-integer quantum Hall effect and Klein tunneling effect, are a direct consequence of this analogical "quasi relativistic" quantum physics. In this lecture, I will make a connection of physics in graphene to relativistic quantum physics employing the concept of pseudo-spin.

Many of the interesting physical phenomena appearing in graphene are governed by the unique chiral nature of the charge carriers in graphene owing to their quasi relativistic quasiparticle dynamics described by the effective massless Dirac equation. This interesting theoretical description can be dated back to Wallace's early work of the electronic band structure calculation of graphite in 1947 where he used the simplest tight binding model and correctly captured the essence of the electronic band structure of graphene, the basic constituent of graphite [1]. Figure 1 shows the reconstructed the band structure of electrons in graphene where the energy can be expressed by 2D momentum in the plane. The valence band (lower band) and conduction band (upper band) touch at six points, where the Fermi level is located. In the vicinity of these points, the energy dispersion relation is linear, mimicking massless particle spectrum in relativistic physics. As we will later discuss in detail, this interesting electronic structures also exhibit the chiral nature of carrier dynamics, another important characteristics of relativistic quantum particles, rediscovered several times in graphene in different contexts [2, 3, 4, 5].

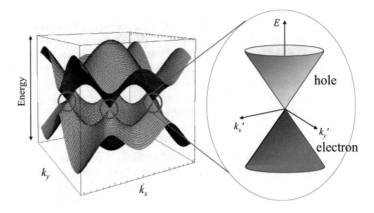

FIGURE 1. Energy band structure of graphene. Linear dispersion relation near the central part of the bands are highlighted. Reproduced from Ref.[10].

In the following sections, after a brief survey of early experiments related to graphene, we will focus on the chiral nature of the electron dynamics in monolayer graphene where the electron wave function's pseudo-spin plays an important role. We will present two experimental examples: the half-integer QHE [6, 7] and the Klein tunneling effect in graphene [8]. The quasi relativistic quantum dynamics of graphene has provided a compact and precise description for these unique experimental observations and further providing a playground for implementing tests of quantum electrodynamics (QED) in a simple experimental situation [9], where electron Fabry–Perrot oscillations were recently observed [8].

2. Early experiment

Independent to earlier theoretical works, experimental efforts to obtain graphene dated back to Böhm et al.'s early work of transmission electron microscopy [11] and the early chemical deposition growth of graphene on metal surfaces developed in the 1990s [12]. In the past decade, renewed efforts to obtain the atomically thin graphite have been pursued through several different routes. In retrospect, these various methods fall into two categories: the bottom-up approach and the top-down approach. In the former, one started with carbon atoms and one tries to assemble graphene sheets by chemical pathways [13, 14]. This is best exemplified by work of the W.A. de Heer group at the Georgia Institute of Technology. In Ref. [14], they demonstrated that thin graphite films can be grown by thermal decomposition on the (0001) surface of 6H-SiC. This method opens the way to large scale integration of nanoelectronics based on graphene. Recent progress through this chemical approach to graphene synthesis has had dazzling successes by diverse routes, including epitaxial graphene growth [15], chemical vapor deposition [16, 17], and solution processing [18].

On the other hand, the top-down approach starts with bulk graphite, which essentially involves graphene sheets stacked together, and tries to extract graphene sheets from the bulk by mechanical exfoliation. The mechanical extraction of layered material dates back to the 1970s. In his seminal experiment [19], Frindt showed that few layers of superconducting $NbSe_2$ can be mechanically cleaved from a bulk crystal fixed on an insulating surface using epoxy. While it is known for decades that people routinely cleave graphite using scotch tape when preparing sample surfaces for Scanning Tunneling Microscopy (STM) study and all optics related studies, the first experiment explicitly involving the mechanical cleavage of graphite using scotch tape was carried out by Ohashi et $al.$ [20]. The thinnest graphite film obtained in this experiment was about 10 nm, corresponding to ∼30 layers.

Experimental work to synthesize very thin graphitic layers directly on top of a substrate [21] or to extract graphene layers using chemical [22] or mechanical [23, 24, 25] exfoliation was demonstrated to produce graphitic samples with thicknesses ranging from 1 to 100 nm. Systematic transport measurements have been carried out on mesoscopic graphitic disks [26] and cleaved bulk crystals [20] with sample thicknesses approaching ∼20 nm, exhibiting mostly bulk graphite properties at these length scales. More controllability was attempted when Ruoff et $al.$ worked out a patterning method for bulk graphite into a mesoscopic scale structure to cleave off thin graphite crystallites using atomic force microscopy [25].

A sudden burst of experimental and theoretical work on graphene followed the first demonstration of single- and multi-layered graphene samples made by a simple mechanical extraction method [27], while several other groups were trying various different routes concurrently [28, 29, 14]. The method that Novoselov et al. used was pretty general, and soon after, it was demonstrated to be applicable to other layered materials [30]. This simple extraction technique is now known as the mechanical exfoliation method. It also has a nick name, "scotch tape" method, since the experimental procedure employs adhesive tapes to cleave off the host crystals before the thin mesocoscopic samples are transferred to a target substrate, often a silicon wafer coated with a thin oxide layer. A carefully tuned oxide thickness is the key to identify single layer graphene samples among the debris of cleaved and transferred mesoscopic graphite samples using the enhanced optical contrast effect due to Fabry–Perot interference [31].

Since this first demonstration of experimental production of an isolated single atomic layer of graphene sample, numerous unique electrical, chemical, and mechanical properties of graphene have been investigated. In particular, an unusual half-integer quantum Hall effect (QHE) and a non-zero Berry phase [6, 7] were discovered in graphene, providing unambiguous evidence for the existence of Dirac fermions in graphene and distinguishing graphene from conventional 2D electronic systems with a finite carrier mass.

3. Pseudospin chirality in graphene

Carbon atoms in graphene are arranged in a honeycomb lattice. This hexagonal arrangement of carbon atoms can be decomposed into two interpenetrating triangular sublattices related to each other by inversion symmetry. Taking two atomic orbitals on each sublattice site as a basis (see Figure 2), the tight binding Hamiltonian can be simplified near two inequivalent Brillouin zone corners, $\mathbf{K}$ and $\mathbf{K}'$) as

$$\hat{H} = \pm \hbar v_F \boldsymbol{\sigma} \cdot (-i\hbar \boldsymbol{\nabla}), \tag{1}$$

where $\boldsymbol{\sigma} = (\sigma_x, \sigma_y)$ are the Pauli matrices, $v_F \approx 10^6$ m/s is the Fermi velocity in graphene and the $+$ $(-)$ sign corresponds to taking the approximation that the wave vector $\mathbf{k}$ is near the $\mathbf{K}$ $(\mathbf{K}')$ point.

The structure of this "Dirac" equation is interesting for several reasons. First, the resulting energy dispersion near the zone corners is linear in momentum, $E(\kappa) = \pm \hbar v_F |\kappa|$, where the wave vector κ is defined relative to $\mathbf{K}$(or $\mathbf{K}'$), i.e.,

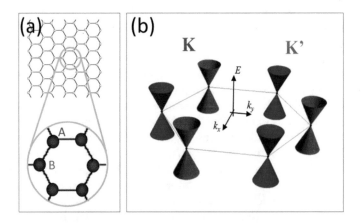

FIGURE 2. (a) Real space image of graphene lattice structure. Two different sub-lattices are marked by red and blue color. (b) Low energy approximation of energy band near the charge neutrality of graphene energy band. Two inequivalent corner of the Brillouin zone are marked by $\mathbf{K}$ and $\mathbf{K}'$, respectively.

$\kappa = \mathbf{k} - \mathbf{K}$(or $\mathbf{K}'$). Consequently, the electrons near these two Dirac points always move at a constant speed, given by the Fermi velocity $v_F \approx c/300$ (rather than the real speed of light c). The electron dynamics in graphene are thus effectively "relativistic", where the speed of light is substituted by the electron Fermi velocity v_F. In a perfect graphene crystal, the Dirac points ($\mathbf{K}$ and $\mathbf{K}'$) are coincident with the overall charge neutrality point (CNP), since there are two carbon atoms in the unit cell of graphene and each carbon atom contributes one electron to the two bands, resulting in the Fermi energy E_F of neutral graphene lying precisely at the half-filled band.

For the Bloch wave function near $\mathbf{K}'$, the "Dirac" equation in equation (1) can be rewritten as

$$\hat{H} = \pm \hbar v_F \boldsymbol{\sigma} \cdot \boldsymbol{\kappa}. \tag{2}$$

The solution of this massless Dirac fermion Hamiltonian is studied by [32, 5, 33]:

$$|\kappa\rangle = \frac{1}{\sqrt{2}} e^{i\boldsymbol{\kappa} \cdot \mathbf{r}} \begin{pmatrix} -ise^{-i\theta_\kappa/2} \\ e^{i\theta_\kappa/2} \end{pmatrix}, \tag{3}$$

where θ_κ is the angle between $\boldsymbol{\kappa} = (\kappa_x, \kappa_y)$ and the y-axis, and $s = +1$ and -1 denote the states above and below $\mathbf{K}$, respectively. The corresponding energy for these states is given by

$$E_s(\kappa) = s\hbar v_F |\boldsymbol{\kappa}|, \tag{4}$$

where $s = +1/-1$ is an index for the positive/negative energy band, respectively. The two components of the state vector give the amplitudes of the electronic wave functions on the atoms of the two sublattices, so the angle θ_κ determines the character of the underlying atomic orbital mixing.

The two-component vector in formula in equation (3) can be viewed as a result of a spinor-rotation of θ_κ around $\hat{z}$-axis with the spin-1/2 rotation operator

$$R(\theta) = \exp\left(-i\frac{\theta}{2}\sigma_z\right) = \begin{pmatrix} e^{-i\theta/2} & 0 \\ 0 & e^{+i\theta/2} \end{pmatrix}. \tag{5}$$

More explicitly, the vectorial part of the Bloch state, $|\mathbf{s}_p\rangle = e^{-i\boldsymbol{\kappa} \cdot \mathbf{r}}|\kappa\rangle$ can be obtained from the initial state along the y-axis,

$$|\mathbf{s}_p^0\rangle = \frac{1}{\sqrt{2}} \begin{pmatrix} -is \\ 1 \end{pmatrix}, \tag{6}$$

by the rotation operation $|\mathbf{s}_p\rangle = R(\theta_\kappa)|\mathbf{s}_p^0\rangle$. Note that this rotation operation clearly resembles that of a two-component spinor describing the electron spin, but arising from the symmetry of the underlying honeycomb graphene lattice. In this regard, $|\mathbf{s}_p\rangle$ is often called "pseudo spin" in contrast to the real spin of electrons in graphene. The above operation also implies that the orientation of the pseudospin is tied to the $\boldsymbol{\kappa}$ vector. This is completely analogous to the real spin of massless fermions which always points along the direction of propagation. For $s = +1$, i.e., corresponding to the upper cone at $\mathbf{K}$ in Figure 1, the states have pseudospin parallel to $\boldsymbol{\kappa}$, and thus correspond to the right-handed Dirac fermions. For $s = -1$, i.e., for the antiparticles in the lower cone, the situation is reversed, resulting in the left-handed Dirac anti-fermions.

So far our analysis is focused on the $\mathbf{K}$ point. It would be interesting to see what happens at the $\mathbf{K}'$ point. We apply a similar analysis at $\mathbf{K}'$, and the only difference is that now we expand the Hamiltonian around the $\mathbf{K}'$ point: $\mathbf{k} = \boldsymbol{\kappa} + \mathbf{K}'$. Then we obtain a new equation for $\mathbf{K}'$ from equation (1),

$$H = \hbar v_F \boldsymbol{\kappa} \cdot \bar{\boldsymbol{\sigma}}, \tag{7}$$

where $\bar{\boldsymbol{\sigma}}$ are the complex conjugate of the Pauli matrices $\boldsymbol{\sigma}$. This Hamiltonian is known to describe left-handed massless neutrinos. Therefore at $\mathbf{K}'$ the electron dynamics is again characterized by massless Dirac fermions, but with opposite helicity.

The chirality of the electrons in graphene has important implications on the electronic transport in graphene. In particular, a non-trivial Berry phase is associated with the rotation of the 1/2-pseudo spinor which plays a critical role to understand the unique charge transport in graphene and nanotubes, as first discussed in Ando et al. [5]'s theoretical work. For example, let us consider a scattering process $\boldsymbol{\kappa} \to \boldsymbol{\kappa}'$ due to a potential $V(\mathbf{r})$ with a range larger than the lattice constant in graphene, so that it does not induce an inter valley scattering between $\mathbf{K}$ and $\mathbf{K}'$ points. The resulting matrix element between these two states is given by [5, 33]

$$|\langle \boldsymbol{\kappa}'|V(\mathbf{r})|\boldsymbol{\kappa}\rangle|^2 = |V(\boldsymbol{\kappa} - \boldsymbol{\kappa}')|^2 \cos^2(\theta_{\kappa,\kappa'}/2), \tag{8}$$

where $\theta_{\kappa,\kappa'}$ is the angle between $\boldsymbol{\kappa}$ and $\boldsymbol{\kappa}'$, and the cosine term comes from the overlap of the initial and final spinors. A backscattering process corresponds to $\boldsymbol{\kappa} = -\boldsymbol{\kappa}'$. In this case, $\theta_{\kappa,\kappa'} = \pi$ and the matrix element vanishes. Therefore such backward scattering is completely suppressed. In terms of the pseudo spin argument, this back scattering process can be described by rotating $|\boldsymbol{\kappa}\rangle$ by the rotating operation $R(\pi)$. For an atomically smooth potential the matrix element in equation (8) can be expressed

$$\langle \boldsymbol{\kappa}'|V(\mathbf{r})|\boldsymbol{\kappa}\rangle \approx V(\boldsymbol{\kappa} - \boldsymbol{\kappa}')\langle \boldsymbol{\kappa}|R(\pi)|\boldsymbol{\kappa}\rangle. \tag{9}$$

Note that a π rotation of the 1/2 spinor always produces an orthogonal spinor to the original one, which makes this matrix element vanish.

The experimental significance of the Berry phase of π was demonstrated by McEuen et al. [33] in single-wall carbon nanotubes (SWCNTs), which are essentially graphene rolled up into cylinders. The suppression of backscattering in metallic SWCNTs leads to a remarkably long electron mean free path on the order of a micron at room temperature [34].

The suppression of backward scattering can also be understood in terms of the Berry phase induced by the pseudo spin rotation. In particular, for complete backscattering, equation (5) yields $R(2\pi) = e^{i\pi}$, indicating that rotation in $\boldsymbol{\kappa}$ by 2π leads to a change of the phase of the wave function $|\boldsymbol{\kappa}\rangle$ by π. This non-trivial Berry phase may lead to non-trivial quantum corrections to the conductivity in graphene, where the quantum correction enhances the classical conductivity. This phenomena is called "anti-localization" in contrast to such quantum corrections in a conventional two-dimensional (2D) system which lead to the suppression of conductivity in a weak localization. This can simply be explained by considering each scattering process with its corresponding complementary time-reversal scattering process. In a conventional 2D electron system such as in GaAs heterojunctions, the scattering amplitude and associated phase of each scattering process and its complementary time-reversal process are equal. This constructive interference in conventional 2D system leads to the enhancement of the backward scattering amplitude and thus results in the localization of the electron states. This mechanism is known as weak localization. In graphene, however, each scattering process and its time reversal pair have a phase difference by π between them due to the non-trivial Berry phase, stemming from 2π rotation of the pseudospin between the scattering processes of the two time reversal pairs. This results in a destructive interference between the time reversal pair to suppress the overal backward scattering amplitude, leading to a positive quantum correction in conductivity. These anti-weak localization phenomena in graphene have been observed experimentally [35].

While the existence of a non-trivial Berry phase in graphene can be inferred indirectly from the aforementioned experiments, it can be directly observed in the quantum oscillations induced by a uniform external magnetic field [6, 7]. In a semi-classical picture, the electrons orbit along a circle in $\mathbf{k}$ space when subjected to a magnetic field. The Berry phase of π produced by the 2π rotation of the wave vector manifests itself as a phase shift of the quantum oscillations, which will be the focus of the discussion of the next section.

4. Berry phase in magneto-oscillations

We now turn to the massless Dirac fermion described by Hamiltonian in equation (1). In a magnetic field, the Schrödinger equation is given by

$$\pm v_F(\mathbf{P} + e\mathbf{A}) \cdot \boldsymbol{\sigma}\psi(\mathbf{r}) = E\psi(\mathbf{r}), \tag{10}$$

where $\mathbf{P} = -i\hbar\triangledown$, $\mathbf{A}$ is the magnetic vector potential, and $\psi(\mathbf{r})$ is a two-component vector

$$\psi(\mathbf{r}) = \begin{pmatrix} \psi_1(\mathbf{r}) \\ \psi_2(\mathbf{r}) \end{pmatrix}. \tag{11}$$

Here we use the Landau gauge $\mathbf{A}$: $\mathbf{A} = -By\hat{x}$ for a constant magnetic field $\mathbf{B}$ perpendicular to the $x - y$ plane. Then, taking only the $+$ sign in equation (10), this equation relates $\psi_1(\mathbf{r})$ and $\psi_2(\mathbf{r})$:

$$v_F(P_x - iP_y - eBy)\psi_2(\mathbf{r}) = E\psi_1(\mathbf{r}), \tag{12}$$

$$v_F(P_x + iP_y - eBy)\psi_1(\mathbf{r}) = E\psi_2(\mathbf{r}). \tag{13}$$

Substituting the first to the second equations above, we obtain the equation for $\psi_2(\mathbf{r})$ only

$$v_F^2(P^2 - 2eByP_x + e^2B^2y^2 - \hbar eB)\psi_2(\mathbf{r}) = E^2\psi_2(\mathbf{r}). \tag{14}$$

The eigenenergies of equation (14) can be found by comparing this equation with a massive carrier Landau system:

$$E_n^2 = 2n\hbar eBv_F^2, \tag{15}$$

where $n = 1, 2, 3 \ldots$ The constant $-\hbar eB$ shifts the LL's by half of the equal spacing between the adjacent LLs, and it also guarantees that there is a LL at $E = 0$, which has the same degeneracy as the other LLs. Putting these expressions together, the eigenenergy for a general LL can be written as [4]

$$E_n = sgn(n)\sqrt{2e\hbar v_F^2|n|B}, \tag{16}$$

where $n > 0$ corresponds to electron-like LLs and $n < 0$ corresponds to hole-like LLs. There is a single LL sitting exactly at $E = 0$, corresponding to $n = 0$, as a result of the chiral symmetry and the particle-hole symmetry.

The square root dependence of the Landau-level energy on n, $E_n \propto \sqrt{n}$, can be understood if we consider the DOS for the relativistic electrons. The linear energy spectrum of 2D massless Dirac fermions implies a linear DOS given by

$$N(E) = \frac{E}{2\pi\hbar^2 v_F^2}. \tag{17}$$

In a magnetic field, the linear DOS collapses into LLs, each of which has the same number of states $2eB/h$. As the energy is increased, there are more states available, so that a smaller spacing between the LLs is needed in order to have the same number of state for each LL. A linear DOS directly results in a square root distribution of the LLs, as shown in Figure 3(c).

A wealth of information can be obtained by measuring the response of the 2D electron system in the presence of a magnetic field. One such measurement is done by passing current through the system and measuring the longitudinal resistivity ρ_{xx}. As we vary the magnetic field, the energies of the LLs change. In particular ρ_{xx} goes through one cycle of oscillations as the Fermi level moves from one LL DOS peak to the next as shown in Figure 3(c). These are the so-called Shubnikov–de Haas (SdH) oscillations. As we note in equation (15), the levels in a 2D massless Dirac fermion system, such as graphene, are shifted by a half-integer relative to the conventional 2D systems, which means that the SdH oscillations will have a phase shift of π, compared with the conventional 2D system.

The phase shift of π is a direct consequence of the Berry phase associated with the massless Dirac fermion in graphene. To further elucidate how the chiral nature of an electron in graphene affects its motion, we resort to a semi-classical model where familiar concepts, such as the electron trajectory, provide us a more intuitive physical picture.

We consider an electron trajectory moving in a plane in a perpendicular magnetic field $\mathbf{B}$. The basic equation for the semi-classical approach is

$$\hbar\dot{\mathbf{k}} = -e(\mathbf{v} \times \mathbf{B}), \tag{18}$$

which simply says that the rate of change of momentum is equal to the Lorentz force. The velocity $\mathbf{v}$ is given by

$$\mathbf{v} = \frac{1}{\hbar} \nabla_{\mathbf{k}} \,\epsilon\,, \tag{19}$$

where ϵ is the energy of the electron. Since the Lorentz force is normal to $\mathbf{v}$, no work is done to the electron and ϵ is a constant of the motion. It immediately follows that electrons move along the orbits given by the intersections of constant energy surfaces with planes perpendicular to the magnetic fields.

Integration of equation (18) with respect to time yields

$$\mathbf{k}(t) - \mathbf{k}(0) = \frac{-eB}{\hbar} \left(\mathbf{R}(t) - \mathbf{R}(0)\right) \times \widehat{\mathbf{B}}, \tag{20}$$

where $\mathbf{R}$ is the position of the electron in real space, and $\widehat{\mathbf{B}}$ is the unit vector along the direction of the magnetic field $\mathbf{B}$. Since the cross product between $\mathbf{R}$ and $\widehat{\mathbf{B}}$ simply rotates $\mathbf{R}$ by $90°$ inside the plane of motion, equation (20) means that the electron trajectory in real space is just its k-space orbit, rotated by $90°$ about $\mathbf{B}$ and scaled by $\hbar/eB$.

It can be further shown that the angular frequency at which the electron moves around the intersection of the constant energy surface is given by

$$\omega_c = \frac{2\pi eB}{\hbar^2} \left(\frac{\partial a_k}{\partial \epsilon}\right)^{-1}, \tag{21}$$

where a_k is the area of the intersection in the k-space. For electrons having an effective mass m^*, we have $\epsilon = \hbar^2 k^2/2m^*$ and a_k is given by $\pi k^2 = 2\pi m^*\epsilon/\hbar^2$, while equation (21) reduces to $\omega_c = eB/m^*$. Comparing this equation with equation (21), we find

$$m^* = \frac{\hbar^2}{2\pi} \left(\frac{\partial a_k}{\partial \epsilon}\right), \tag{22}$$

which is actually the definition of the effective mass for an arbitrary orbit.

The quantization of the electron motion will restrict the available states and will give rise to quantum oscillations such as SdH oscillations. The Bohr–Sommerfeld quantization rule for a periodic motion is

$$\oint \mathbf{p} \cdot d\mathbf{q} = (n + \gamma)2\pi\hbar, \tag{23}$$

where $\mathbf{p}$ and $\mathbf{q}$ are canonically conjugate variables, n is an integer and the integration in equation (23) is for a complete orbit. The quantity γ will be discussed below.

For an electron in a magnetic field,

$$\mathbf{p} = \hbar\mathbf{k} - e\mathbf{A}, \qquad \mathbf{q} = \mathbf{R}, \tag{24}$$

so equation (23) becomes

$$\oint (\hbar \mathbf{k} - e\mathbf{A}) \cdot d\mathbf{R} = (n + \gamma)2\pi\hbar. \tag{25}$$

Substituting this equation into equation (20) and using Stokes' theorem, one finds

$$\mathbf{B} \cdot \oint \mathbf{R} \times d\mathbf{R} - \int_S \mathbf{B} \cdot d\mathbf{S} = (n + \gamma)\Phi_0, \tag{26}$$

where $\Phi_0 = 2\pi\hbar/e$ is the magnetic flux quanta. S is any surface in real space which has the electron orbit as the projection on the plane. Therefore the second term on the left-hand side of equation (26) is just the magnetic flux $-\Phi$ penetrating the electron orbit. A closer inspection of the first term on the left-hand side of equation (26) finds that it is 2Φ. Putting them together, equation (26) reduces to

$$\Phi = (n + \gamma)\Phi_0, \tag{27}$$

which simply means that the quantization rule dictates that the magnetic flux through the electron orbit has to be quantized.

Remember that the electron trajectory in real space is just a rotated version of its trajectory in k-space, scaled by $\hbar/eB$ (equation (20)). Let $a_k(\epsilon)$ be the area of the electron orbit at constant energy ϵ in k-space; then equation (20) becomes

$$a_k(\epsilon_n) = (n + \gamma)2\pi eB/\hbar, \tag{28}$$

which is the famous Onsager relation. This relation implicitly specifies the permitted energy levels ϵ_n (Landau levels), which in general depend on the band structure dispersion relation $\epsilon(k)$.

The dimensionless parameter $0 \leq \gamma < 1$ is determined by the shape of the energy band structure. For a parabolic band, $\epsilon = \hbar^2 k^2/2m^*$, the nth LL has the energy $\epsilon_n = (n + 1/2)\hbar\omega_c$. Each LL orbit for an isotropic m^* in the plane perpendicular to the magnetic field B is a circle in k-space with a radius $k_n = \sqrt{2eB(n + 1/2)/\hbar}$. The corresponding area of an orbital in k-space for the nth LL is therefore

$$a_k(\epsilon_n) = \pi k_n^2 = (n + \frac{1}{2})2\pi eB/\hbar. \tag{29}$$

A comparison of this formula with equation (28) immediately yields

$$\gamma = \frac{1}{2}. \tag{30}$$

For a massless Dirac fermion in graphene which obeys a linear dispersion relation $\epsilon = \hbar v_F k$, the nth LL corresponds to a circular orbit with radius $k_n = \epsilon_n/\hbar v_F = \sqrt{2e|n|B/\hbar}$. The corresponding area is therefore

$$a_k(\epsilon_n) = \pi k_n^2 = |n|2\pi eB/\hbar. \tag{31}$$

This gives, for a semiclassical Shubnikov–de Haas (SdH) phase,

$$\gamma = 0, \tag{32}$$

which differs from the γ for the conventional massive fermion by $\frac{1}{2}$.

The difference of $\frac{1}{2}$ in γ is a consequence of the chiral nature of the massless Dirac fermions in graphene. An electron in graphene always has the pseudospin $|s_p\rangle$ tied to its wave vector $\mathbf{k}$. The electron goes through the orbit for one cycle, $\mathbf{k}$, as well as the pseudospin attached to the electrons. Both go through a rotation of 2π at the same time. Much like a physical spin, a 2π adiabatic rotation of pseudospin gives a Berry phase of π [5]. This is exactly where the $\frac{1}{2}$ difference in γ comes from.

The above analysis can be generalized to systems with an arbitrary band structure. In general, γ can be expressed in terms of the Berry phase, ϕ_B, for the electron orbit:

$$\gamma - \frac{1}{2} = -\frac{1}{2\pi}\phi_B . \tag{33}$$

For any electron orbits which surround a disconnected electronic energy band, as is the case for a parabolic band, this phase is zero, and we arrive at equation (30). A non-trivial Berry phase of π results if the orbit surrounds a contact between the bands, and the energies of the bands separate linearly in $\mathbf{k}$ in the vicinity of the band contact. In monolayer graphene, these requirements are fulfilled because the valence band and conduction band are connected by $\mathbf{K}$ and $\mathbf{K}'$, and the energy dispersion is linear around these points. This special situation again leads to a $\gamma = 0$ in graphene (in fact, $\gamma = 0$ and $\gamma = \pm 1$ are equivalent). Note that this non-trivial γ is only for monolayer graphene; in contrast for bilayer graphene, whose band contact points at the charge neutrality point have a quadratic dispersion relation, a conventional $\gamma = 1/2$ is obtained.

γ can be probed experimentally by measuring the quantum oscillation of the 2D system in the presence of a magnetic field, where γ is identified as the phase of such oscillations. This becomes evident when we explicitly write the oscillatory part of the quantum oscillations, e.g., as for the SdH oscillation of the electrical resistivity, $\Delta\rho_{xx}$ [7]

$$\Delta\rho_{xx} = R(B,T)\cos\left[2\pi\left(\frac{B_F}{B} - \gamma\right)\right]. \tag{34}$$

Here we only take account of the first harmonic, in which $R(B,T)$ is the amplitude of the SdH oscillations and B_F is the frequency in units of $1/B$, which can be related to the 2D charge carrier density n_s by

$$B_F = \frac{n_s h}{g_s e} , \tag{35}$$

where $g_s = 4$ accounts for the spin and valley degeneracies of the LLs. The relation $\gamma = 1/2$ (for a parabolic band) and $\gamma = 0$ (for graphene) produces a phase difference of π between the SdH oscillation in the two types of 2D systems. In the extreme quantum limit, the SdH oscillations evolve into the quantum Hall effect, a remarkable macroscopic quantum phenomenon characterized by a precisely quantized Hall resistance and zeros in the longitudinal magneto-resistance. The additional Berry phase of π manifests itself as a half-integer shift in the quantization condition, and leads to an unconventional quantum Hall effect. In the quantum

Hall regime, graphene thus exhibits a so-called "half-integer" shifted quantum Hall effect, where the filling fraction is given by $\nu = g_s(n + 1/2)$ for integer n. Thus, at this filling fraction $\rho_{xx} = 0$, while the Hall resistivity exhibits quantized plateaus at

$$\rho_{xy}^{-1} = \frac{e^2}{h} g_s \left(n + \frac{1}{2} \right). \tag{36}$$

The experimental observation of the quantum Hall effect and Berry phase in graphene were first reported in Novoselov et al. [6] and Zhang et al. [7]. Figure 3a shows R_{xy} and R_{xx} of a single layer graphene sample as a function of magnetic field B at a fixed gate voltage $V_g > V_{\text{Dirac}}$. The overall positive R_{xy} indicates that the contribution to R_{xy} is mainly from electrons. At high magnetic field, $R_{xy}(B)$ exhibits plateaus and R_{xx} is vanishing, which is the hallmark of the QHE. At least two well-defined plateaus with values $(2e^2/h)^{-1}$ and $(6e^2/h)^{-1}$, followed by a developing $(10e^2/h)^{-1}$ plateau, are observed before the QHE features are transformed into Shubnikov–de Haas (SdH) oscillations at lower magnetic field. The quantization of R_{xy} for these first two plateaus is better than 1 part in 10^4, with a precision within the instrumental uncertainty. In recent experiments [36], this limit is now at the accuracy of the 10^{-9} level. We observe the equivalent QHE features for holes ($V_g < V_{\text{Dirac}}$) with negative R_{xy} values (Figure 3a, inset). Alternatively, we can probe the QHE in both electrons and holes by fixing the magnetic field and by changing V_g across the Dirac point. In this case, as V_g increases, first holes ($V_g < V_{\text{Dirac}}$) and later electrons ($V_g > V_{rmDirac}$) are filling successive Landau levels and thereby exhibit the QHE. This yields an antisymmetric (symmetric) pattern of R_{xy} (R_{xx}) in Figure 3b, with R_{xy} quantization accordance to

$$R_{xy}^{-1} = \pm g_s \left(n + \frac{1}{2} \right) e^2/h, \tag{37}$$

where n is a non-negative integer, and +/- stands for electrons and holes, respectively. This quantization condition can be translated to the quantized filling factor, ν, in the usual QHE language. Here in the case of graphene, $g_s = 4$, accounting for 2 by the spin degeneracy and 2 by the sub-lattice degeneracy, equivalent to the $\mathbf{K}$ and $\mathbf{K}'$ valley degeneracy under a magnetic field.

The observed QHE in graphene is distinctively different from the "conventional" QHEs because of the additional half-integer in the quantization condition (equation (37)). This unusual quantization condition is a result of the topologically exceptional electronic structure of graphene. The sequence of half-integer multiples of quantum Hall plateaus has been predicted by several theories which combine "relativistic" Landau levels with the particle-hole symmetry of graphene [37, 38, 39]. This can be easily understood from the calculated LL spectrum (equation (16)) as shown in Figure 3(c). Here we plot the density of states (DOS) of the g_s-fold degenerate (spin and sublattice) LLs and the corresponding Hall conductance ($\sigma_{xy} = -R_{xy}^{-1}$, for $R_{xx} \to 0$) in the quantum Hall regime as a function of energy. Here σ_{xy} exhibits QHE plateaus when E_F (tuned by V_g) falls between LLs, and jumps by an amount of $g_s e^2/h$ when E_F crosses a LL. Time reversal

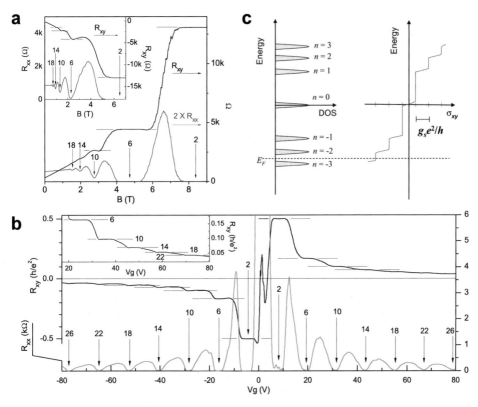

FIGURE 3. Quantized magnetoresistance and Hall resistance of a graphene device. (a) Hall resistance (black) and magnetoresistance (red) measured in a monolayer graphene device at $T = 30$ mK and $V_g = 15$ V. The vertical arrows and the numbers on them indicate the values of B and the corresponding filling factor ν of the quantum Hall states. The horizontal lines correspond to $h/\nu e^2$ values. The QHE in the electron gas is demonstrated by at least two quantized plateaus in R_{xy} with vanishing R_{xx} in the corresponding magnetic field regime. The inset shows the QHE for a hole gas at $V_g = -4$ V, measured at 1.6 K. The quantized plateau for filling factor $\nu = 2$ is well defined and the second and the third plateaus with $\nu = 6$ and 10 are also resolved. (b) The Hall resistance (black) and magnetoresistance (orange) as a function of gate voltage at fixed magnetic field $B = 9$ T, measured at 1.6 K. The same convention as in (a) is used here. The upper inset shows a detailed view of high filling factor R_{xy} plateaus measured at 30 mK. (c) A schematic diagram of the Landau-level density of states (DOS) and corresponding quantum Hall conductance (σ_{xy}) as a function of energy. Note that in the quantum Hall states, $\sigma_{xy} = -R_{xy}^{-1}$. The LL index n is shown next to the DOS peak. In our experiment, the Fermi energy E_F can be adjusted by the gate voltage, and R_{xy}^{-1} changes by an amount of $g_s e^2/h$ as E_F crosses a LL. Reproduced from Ref. [7].

invariance guarantees particle-hole symmetry and thus σ_{xy} is an odd function in energy across the Dirac point [4]. However, in graphene, the $n = 0$ LL is robust, i.e., $E_0 = 0$ regardless of the magnetic field, provided that the sublattice symmetry is preserved [4]. Thus the first plateaus of R_{xy}^{-1} for electrons ($n = 1$) and holes ($n = -1$) are situated exactly at $g_s e^2/2h$. As E_F crosses the next electron (hole) LL, R_{xy}^{-1} increases (decreases) by an amount of $g_s e^2/h$, which yields the quantization condition in equation (37).

A consequence of the combination of time reversal symmetry with the novel Dirac point structure can be viewed in terms of Berry phase arising from the band degeneracy point. A direct implication of Berry's phase in graphene is discussed in the context of the quantum phase of a spin-1/2 pseudo-spinor that describes the sublattice symmetry [5]. This phase is already implicit in the half-integer shifted quantization rules of the QHE.

The linear energy dispersion relation also leads to a linearly vanishing 2D density of states near the charge neutrality point (CNP) at $E = 0$, $\rho_{2D} \propto |\epsilon_F|$. This differs from that for conventional parabolic 2D systems in which the density of states, at least in the single particle picture, is constant, leading to a decrease in the ability of charge neutral graphene to screen electric fields. Finally, the sublattice symmetry endows the quasiparticles with a conserved quantum number and chirality, corresponding to the projection of the pseudospin on the direction of motion [9]. In the absence of scattering which mixes the electrons in the graphene valleys, pseudospin conservation forbids backscattering in graphene [5], momentum reversal being equivalent to the violation of pseudospin conservation. This absence of backscattering has been advanced as an explanation for the experimentally observed unusually long mean free path of carriers in metallic as compared with semiconducting nanotubes [33].

5. Pseudospin and Klein tunneling in graphene

The observation of electron and hole puddles in charge neutral, substrate supported graphene confirmed theoretical expectations [42] that transport at charge neutrality is dominated by charged impurity-induced inhomogeneities. The picture of transport at the Dirac point is as a result of conducting puddles separated by a network of p-n junctions. Understanding the properties of graphene p-n junctions is thus crucial to quantitative understanding of the minimal conductivity, a problem that has intrigued experimentalists and theorists alike [6, 42]. Describing transport in the inhomogeneous potential landscape of the CNP requires introduction of an additional spatially varying electrical potential into equation (1) in the previous section; transport across a p-n junction corresponds to this varying potential crossing zero. Because graphene carriers have no mass, graphene p-n junctions provide a condensed matter analogue of the so-called "Klein tunneling" problem in quantum electro-dynamics (QED). The first part of this section will be devoted to the theoretical understanding of ballistic and diffusive transport across such as barrier.

In recent years, substantial effort has been devoted to improving graphene sample quality by eliminating unintentional inhomogeneity. Some progress in this direction has been made by both suspending graphene samples [43, 44] as well as by transferring graphene samples to single crystal hexagonal boron nitride substrates [45]. These techniques have succeeded in lowering the residual charge density present at charge neutrality, but even the cleanest samples are not ballistic on length scales comparable to the sample size (typically $\gtrsim 1~\mu$m). An alterative approach is to try to restrict the region of interest being studied by the use of local gates.

Graphene's gapless spectrum allows the fabrication of adjacent regions of positive and negative doping through the use of local electrostatic gates. Such heterojunctions offer a simple arena in which to study the peculiar properties of graphene's massless Dirac charge carriers, including chirality [46, 9] and emergent Lorentz invariance [48]. Technologically, graphene p-n junctions are relevant for various electronic devices, including applications in conventional analog and digital circuits as well as novel electronic devices based on electronic lensing. In the latter part of this review, we will discuss current experimental progress towards such gate-engineered coherent quantum graphene devices.

The approach outlined in the previous section requires only small modifications to apply the approach to the case of carrier transport across graphene heterojunctions. While the direct calculation for the case of graphene was done by Katsnelson et al. [9], a similar approach taking into account the chiral nature of carriers was already discussed a decade ago in the context of electrical conduction in metallic carbon nanotubes [5]. In low-dimensional graphitic systems, the free particle states described by equation (1) are chiral, meaning that their pseudospin is parallel (antiparallel) to their momentum for electrons (holes). This causes a suppression of backscattering in the absence of pseudospin-flip nonconserving processes, leading to the higher conductances of metallic over semiconducting carbon nanotubes [33]. To understand the interplay between this effect and Klein tunneling in graphene, we introduce external potentials $\boldsymbol{A}(\boldsymbol{r})$ and $U(\boldsymbol{r})$ in the Dirac Hamiltonian,

$$\hat{H} = v_F \boldsymbol{\sigma} \cdot (-i\hbar \boldsymbol{\nabla} - e\boldsymbol{A}(\boldsymbol{r})) + U(\boldsymbol{r}). \tag{38}$$

In the case of a one-dimensional (1D) barrier, $U(\boldsymbol{r}) = U(x)$, at zero magnetic field, and the momentum component parallel to the barrier, p_y, is conserved. As a result, electrons normally incident on a graphene p-n junction are forbidden from scattering obliquely by the symmetry of the potential, while chirality forbids them from scattering directly backwards: the result is perfect transmission as holes [9], and this is what is meant by Klein tunneling in graphene (see (Figure 4) (a)). The rest of this review is concerned with gate induced p-n junctions in graphene; however, the necessarily transmissive nature of graphene p-n junctions is crucial for understanding the minimal conductivity and supercritical Coulomb impurity problems in graphene, as well as playing a role in efforts to confine graphene quantum particles. Moreover, p-n junctions appear in the normal process of contacting and locally gating graphene, both of which are indispensable for electronics applications.

Even in graphene, an atomically sharp potential cannot be created in a realistic sample. Usually, the distance to the local gate, which is isolated from the graphene by a thin dielectric layer determines the length scale on which the potential varies. The resulting transmission problem over a Sauter-like potential step in graphene was solved by Cheianov and Fal'ko [46]. Substituting the Fermi energy for the potential energy difference $\varepsilon - U(x) = \hbar v_f k_f(x)$ and taking into account the conservation of the momentum component $p_y = \hbar k_F \sin\theta$ parallel to the barrier, they obtained a result, valid for $\theta \ll \pi/2$, that is nearly identical to that of Sauter [47]:

$$
k_F(x) = \begin{cases} -k_F/2 & x < 0, \\ Fx & 0 \le x \le L, \\ k_F/2 & x > L. \end{cases} \qquad |T|^2 \sim \exp\left(-2\pi^2 \frac{\hbar v_F}{F\lambda_F^2}\sin^2\theta\right), \qquad (39)
$$

As in the massive relativistic problem in one-dimension, the transmission is determined by evanescent transport in classically forbidden regions where $k_x(x)^2 = k_F(x)^2 - p_y^2 < 0$ (Figure 4). The only differences between the graphene case and the one-dimensional, massive relativistic case are the replacement of the speed of light by the graphene Fermi velocity, the replacement of the Compton wavelength by the Fermi wavelength, and the scaling of the mass appearing in the transmission by the sine of the incident angle. By considering different angles of transmission in the barrier problem in two-dimensional graphene, then, one can access both the Klein and Sauter regimes of $T \sim 1$ and $T \ll 1$.

The current state of the experimental art in graphene does not allow for injection of electrons with definite p_y [8, 49]. Instead, electrons impinging on a p-n junction have a random distribution of incident angles due to scattering in the diffusive graphene leads. Equation 39 implies that in realistically sharp p-n junctions, these randomly incident electrons emerge from the p-n junction as a collimated beam, with most off-normally incident carriers being scattered; transmission through multiple p-n junctions leads to further collimation. Importantly, even in clean graphene, taking into account the finite slope of the barrier yields qualitatively different results for the transmission: just as in the original Klein problem, the sharp potential step [9] introduces pathologies – in the case of graphene, high transmission at $\theta \ne 0$ – which disappear in the more realistic treatment [46].

Transport measurements across single p-n junctions, or a pnp junction in which transport is not coherent, can at best provide only indirect evidence for Klein tunneling by comparison of the measured resistance of the p-n junction. Moreover, because such experiments probe only incident-angle averaged transmission, they cannot experimentally probe the structure $T(\theta)$. Thus, although the resistance of nearly ballistic p-n junctions are in agreement with the ballistic theory, to show that angular collimation occurs, or that there is perfect transmission at normal incidence, requires a different experiment. In particular, there is no way to distinguish perfect transmission at $\theta = 0$ from large transmission at all

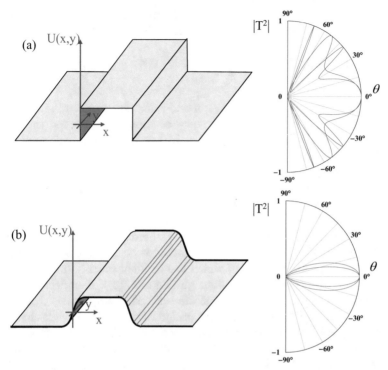

FIGURE 4. Potential landscape and angular dependence of quasiparticle transmission through (a) an atomically sharp pnp barrier and (b) an electrostatically generated smooth pnp barrier in graphene, with their respective angle-dependent transmission probabilities $|T|^2$. Red and blue lines correspond to different densities in the locally gated region.

angles, begging the question of whether "Klein tunneling" has any observable consequences outside the context of an angle resolved measurement or its contribution to bulk properties such as the minimal conductivity. In fact, as was pointed out by Shytov et al. [50], an experimental signature of this phenomenon should manifest itself as a sudden phase shift at finite magnetic field in the transmission resonances in a ballistic, phase coherent, graphene pnp device.

Although graphene p-n junctions are transmissive when compared with p-n junctions in gapfull materials (or gapless materials in which backscattering is allowed, such as bilayer graphene), graphene p-n junctions are sufficiently reflective, particularly for obliquely incident carriers, to cause transmission resonances due to Fabry–Perot interference. However, in contrast to the canonical example from optics, or to one-dimensional electronic analogues, the relative phase of interfering paths in a ballistic, phase coherent pnp (or npn) graphene heterojunction can be tuned by applying a magnetic field. For the case where the junction width is only somewhat shorter than the mean free path in the local gate region (LGR),

$L \lesssim \ell_{LGR}$, the Landau formula for the oscillating part of the conductance trace can be derived from the ray tracing diagrams in Figure 5(d),

$$G_{osc} = e^{-2L/l_{LGR}} \frac{4e^2}{h} \sum_{k_y} 2|T_+|^2 |T_-|^2 R_+ R_- \cos(\theta_{WKB}), \quad (40)$$

in which $T_\pm$ and $R_\pm$ are the transmission and reflection amplitudes at $x = \pm L/2$, θ_{WKB} is the semiclassical phase difference accumulated between the junctions by interfering trajectories, and ℓ_{LGR} is a fitting parameter which controls the amplitude of the oscillations.

At zero magnetic field, particles are incident at the same angle on both junctions, and the Landau sum in equation (40) is dominated by the modes for which both transmission and reflection are nonnegligible, so neither normal nor highly oblique modes contribute. Instead, the sum is dominated by modes with finite k_y, peaked about $k_y = \pm\sqrt{F/(\ln(3/2)\pi\hbar v_F)}$, where F is the electric force in the pn junction region. As the magnetic field increases, cyclotron bending favors the contribution of modes with $k_y = 0$, which are incident on the junctions at angles with the same magnitude but opposite sign (Figure 5(c)). If perfect transmission at zero angle exists, then analyticity of the scattering amplitudes demands that the reflection amplitude changes sign as the sign of the incident angle changes [50], thereby causing a π shift in the reflection phase. This effect can also be described in terms of the Berry phase: the closed momentum space trajectories of the modes dominating the sum at low field and high k_y do not enclose the origin, while those at intermediate magnetic fields and $k_y \sim 0$ do. As a consequence, the quantization condition leading to transmission resonances is different due to the inclusion of the Berry phase when the trajectories surround the topological singularity at the origin, leading to a phase shift in the observed conductance oscillations as the phase shift containing trajectories begin to dominate the Landau sum in equation (40).

Experimental realization of the coherent electron transport in pnp (as well as npn) graphene heterojunctions was reported by Young and Kim [8]. The key experimental innovations were to use an extremely narrow ($\lesssim 20$ nm wide) top gate, creating a Fabry–Perot cavity between p-n junctions smaller than the mean free path, which was ~ 100 nm in the samples studied. Figure 5(a) shows the layout of a graphene heterojunction device controlled by both top gate voltage (V_{TG}) and back gate voltage (V_{BG}). The conductance map shows clear periodic features in the presence of p-n junctions; these features appear as oscillatory features in the conductance as a function of V_{TG} at fixed V_{BG} (Figure 5(b)). For the electrostatics of the devices presented in this device, the magnetic field at which this π phase shift, due to the Berry phase at the critical magnetic field discussed above, is expected to occur in the range $B^* = 2\hbar k_y/eL \sim 250$–500 mT, in agreement with experimental data which show an abrupt phase shift in the oscillations at a few hundred mT (Figure 5(f)). Experiments can be matched quantitatively to the theory by calculation of equation (40) for the appropriate potential profile, providing confirmation of the Klein tunneling phenomenon in graphene. As the magnetic field increases further, the ballistic theory predicts the disappearance of the Fabry–Perot conductance oscillations as the cyclotron radius shrinks below the distance between p-n junctions, $R_c \lesssim L$, or B$\sim$2 T for our devices (Figure 5(e)). The qualitative understanding of these behaviors can be explained in the semiclassical way (Figure 5(c)): with increasing B, the dominant modes at low magnetic field (blue) give way to phase-shifted modes with negative reflection amplitude due to the inclusion of the non-trivial Berry phase (orange), near $k_y = 0$. The original finite k_y modes are not yet phase shifted at the critical magnetic field B$_c$, above which the non-trivial Berry phase shift π (green) appears. But owing to collimation, these finite k_y modes no longer contribute to the oscillatory conductance.

There is an apparent continuation of the low magnetic field Fabry–Perot (FP) oscillations to Shubnikov–de Haas (SdH) oscillations at high magnetic fields. Generally, the FP oscillations tend to be suppressed at high magnetic fields as the cyclotron orbits get smaller than the junction size. On the other hand, disorder mediated SdH oscillations become stronger at high magnetic field owing to the large separation between Landau levels. The observed smooth continuation between these two oscillations does not occur by chance. FP oscillations at magnetic fields higher than the phase shift are dominated by trajectories with $k_y = 0$; similarly, SdH oscillations, which can be envisioned as cyclotron orbits beginning and ending on the same impurity, must also be dominated by $k_y = 0$ trajectories. The result is a seamless crossover from FP to SdH oscillations. This is strongly dependent on the disorder concentration: for zero disorder, SdH oscillations do not occur, while for very strong disorder SdH oscillations happen only at high fields and FP oscillations do not occur due to scattering between the p-n junctions. For low values of disorder, such that SdH oscillations appear at fields much smaller than the phase shift magnetic field, B$_c$, the two types of coherent oscillations could in principle coexist with different phases. The role of disorder in the FP-SdH crossover has only begun to be addressed experimentally [8].

6. Conclusions

In this chapter, we discuss the role of pseudospin in electronic transport in graphene. We demonstrate a variety of new phenomena which stem from the effectively relativistic nature of the electron dynamics in graphene, where the pseudospin is aligned with two-dimensional momentum. Our main focus were two major topics: (i) the non-conventional, half-integer shifted filling factors for the quantum Hall effect (QHE) and the peculiar magneto-oscillation where one can directly probe the existence of a non-trivial Berry phase, and (ii) Klein tunneling of chiral Dirac fermions in a graphene lateral heterojunction.

Employing unusual filling factors in QHE in single layer graphene samples as an example, we demonstrated that the observed quantization condition in graphene is described by half-integer rather than integer values, indicating the contribution

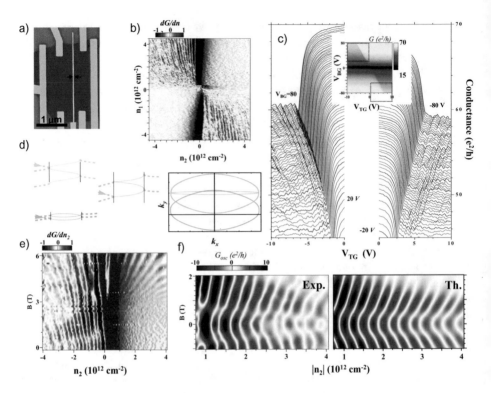

FIGURE 5. (a) Scanning electron microscope image of a typical graphene hetero-junction device. Electrodes, graphene and top gates are represented by yellow, purple and cyan, respectively. (b) A differential transconductance map of the device as a function of densities n_2 and n_1, corresponding to the locally gated region (LGR) and out side the LGR, i.e., graphene lead (GL) region, respectively. Interference fringes appear in the presence of pn junctions, which define the Fabry–Perot cavity. (c) Inset: Conductance map of the device in the back gate and top gate voltages (V_{BG}-V_{TG}) plane. The main panels show cuts through this color map in the regions indicated by the dotted lines in the inset, showing the conductance as a function of V_{TG} at fixed V_{BG}. Traces are separated by a step in V_{BG} of 1 V, starting from 80V with traces taken at integer multiples of 5 V in black. (d) Schematic diagram of trajectories contributing to quantum oscillations in real and momentum space. (e) Magnetic-field and density dependence of the transconductance dG/dn_2 for $n_1 > 0$ is fixed. Note that the low field oscillatory features from FP resonance only appear for $n_2 < 0$ where there is pnp junction forms. (f) Oscillating part of the conductance at V_{BG} =50 V for low fields. G_{osc} as extracted from the experimental data over a wide range of densities and magnetic fields (left) matches the behavior predicted by a theory containing the phase shift due to Klein tunneling [50] (right). Reproduced from Ref. [8].

of the non-trivial Berry phase. The half-integer quantization, as well as the measured phase shift in the observed magneto-oscillations, can be attributed to the peculiar topology of the graphene band structure with a linear dispersion relation and a vanishing mass near the Dirac point, which is described in terms of effectively "relativistic" carriers as shown in equation (1). The unique behavior of electrons in this newly discovered $(2 + 1)$-dimensional quantum electrodynamics system not only opens up many interesting questions in mesoscopic transport in electronic systems with non-zero Berry phase but may also provide the basis for novel carbon based electronics applications.

The development and current status of electron transport in graphene heterojunction structures were also reviewed. In these lateral heterojunction devices, the unique linear energy dispersion relation and concomitant pseudospin symmetry are probed via the use of local electrostatic gates. Mimicking relativistic quantum particle dynamics, electron waves passing between two regions of graphene with different carrier densities will undergo strong refraction at the interface, producing an experimental realization of the century-old Klein tunneling problem of relativistic quantum mechanics. Many theoretical and experimental discussions were presented here, including the peculiar graphene p-n and *pnp* junction conduction in the diffusive and ballistic regimes. Since electrons are charged, a magnetic field can couple to them and magnetic field effects can be studied. In particular, in a coherent system, the electron waves can also interfere, producing quantum oscillations in the electrical conductance, which can be controlled through the application of both electric and magnetic fields. A clear indication of a phase shift of π in the magnetoconductance clearly indicates again the existence of a non-trivial Berry phase associated with the pseudo-spin rotation during the Klein tunneling process.

References

[1] Wallace, P.R.: Phys. Rev. **71**, 622 (1947).

[2] DiVincenzo, D.P., and Mele, E.J.: Phys. Rev. B **29**, 1685 (1984).

[3] Semenoff, G.W.: Phys. Rev. Lett. **53**, 2449 (1984).

[4] Haldane, F.D.M.: Phys. Rev. Lett. **61**, 2015 (1988).

[5] Ando, T., Nakanishi, T., and Saito, R.: J. Phys. Soc. Jpn **67**, 2857 (1998).

[6] Novoselov, K.S., Geim, A.K., Morozov, S.V., Jiang, D., Katsnelson, M. I., Grigorieva, I.V., Dubonos, S.V., and Firsov, A.A.: Nature **438**, 197 (2005).

[7] Zhang, Y., Tan, Y.-W., Stormer, H.L., and Kim, P.: Nature **438**, 201 (2005).

[8] Young, A.F., and Kim, P.: Nat. Phys. **5**, 222 (2009).

[9] Katsnelson, M.I., Novoselov, K.S., and Geim, A.K.: Nat. Phys. **2**, 620 (2006).

[10] Wilson, M.: Physics Today **1**, 21 (2006).

[11] Boehm, H.P., Clauss, A., Fischer, G.O., Hofmann, U.: Z. anorg. allg. Chem. **316**, 119 (1962).

[12] Land, T.A., Michely, T., Behm, R.J., Hemminger, J.C., Comsa, G.: Sur. Sci. **264**, 261–270 (1992).

[13] Krishnan, A., Dujardin, E., Treacy, M.M.J., Hugdahl, J., Lynum, S., and Ebbesen, T.W.: Nature **388**, 451 (1997).

[14] Berger, C., Song, Z.M., Li, T.B., Li, X.B., Ogbazghi, A.Y., Feng, R., Dai, Z.T., Marchenkov, A.N., Conrad, E.H., First, P.N., and de Heer, W.A.: J. Phys. Chem. B **108**, 19912 (2004).

[15] de Heer, W.A., Berger, C., Ruan, M., Sprinkle, M., Li, X., Yike Hu, Zhang, B., Hankinson, J., Conrad, E.H.: PNAS **108**, (41) 16900 (2011).

[16] Bae, S. *et al.*: Nature Nanotech. **5**, 574 (2010).

[17] Yamamoto, M., Obata, S., and Saiki, K.: Surface and Interface Analysis **42**, 1637 (2010).

[18] Zhang, L. *et al.*: Nano Lett. **12**, 1806 (2012).

[19] Frindt, R.F.: Phys. Rev. Lett. **28**, 299 (1972).

[20] Ohashi, Y., Hironaka, T., Kubo, T., and Shiki, K.: Tanso 1997, 235 (1997).

[21] Itoh, H., Ichinose, T., Oshima, C., and Ichinokawa, T.: Surf. Sci. Lett. **254**, L437 (1991).

[22] Viculis, L.M., Jack, J.J., and Kaner, R.B.: Science **299**, 1361 (2003).

[23] Ebbensen, W., and Hiura, H.: Adv. Mater. **7**, 582 (1995).

[24] Lu, X., Huang, H., Nemchuk, N., and Ruoff, R.: Appl. Phys. Lett. **75**, 193 (1999).

[25] Lu, X., Yu, M., Huang, H., and Ruoff, R.: Nanotechnology **10**, 269 (1999).

[26] Dujardin, E., Thio, T., Lezec, H., and Ebbesen, T.W.: Appl. Phys. Lett. **79**, 2474 (2001).

[27] Novoselov, K.S., Geim1, A.K., Morozov, S.V., Jiang, D., Zhang, Y., Dubonos, S.V., Grigorieva, I.V., Firsov, A.A.: Science **306**, 666 (2004).

[28] Zhang, Y., Small, J.P., Pontius, W.V., and Kim, P.: App. Phys. Lett. **86**, 073104 (2005).

[29] Bunch, J.S., Yaish, Y., Brink, M., Bolotin, K., and McEuen, P.L.: Nano Lett. **5**, 287 (2005).

[30] Novoselov, K.S., Jiang, D., Schedin, F., Booth, T.J., Khotkevich, V.V., Morozov, S.V., and Geim, A.K.: PNAS **102**, 10451 (2005).

[31] Blake, P., Hill, E.W., Castro Neto, A.H., Novoselov, K.S., Jiang, D., Yang, R., Booth, T.J., and Geim, A.K.: Appl. Phys. Lett. **91**, 063124 (2007).

[32] Slonczewski, J.C., and Weiss, P.R.: Phys. Rev. **109**, 272 (1958).

[33] McEuen, P.L., Bockrath, M., Cobden, D.H., Yoon, Y.-G., and Louie, S.G.: Phys. Rev. Lett. **83**, 5098 (1999).

[34] Purewal, M.S., Hong, B.H., Ravi, A., Chandra, B., Hone, J., and Kim, P.: Phys. Rev. Lett. **98**, 196808 (2006).

[35] Tikhonenko, F.V., Horsell, D.W., Gorbachev, R.V., and Savchenko, A.K.: Phys. Rev. Lett. **100**, 056802 (2008).

[36] Tzalenchuk1, A., Lara-Avila, S., Kalaboukhov, A., Paolillo, S., Syvajarvi, M., Yakimova, R., Kazakova, O., Janssen, T.J.B.M., Fal'ko, V., and Kubatkin, S.: Nature Nano. **5**, 186 (2010).

[37] Zheng, Y.S., and Ando, T.: Phys. Rev. B **65**, 245420 (2002).

[38] Gusynin, V.P., and Sharapov, S.G.: Phys. Rev. Lett. **95**, 146801 (2005).

[39] Peres, N.M.R., Guinea, F., and Neto, A.H.C.: Phys. Rev. B **73**, 125411 (2006).

[40] Morozov, S.V., Novoselov, K.S., Schedin, F., Jiang, D., Firsov, A.A., and Geim, A.K.: Phys. Rev. B **72**, 201401 (R) (2005).

[41] Zhang, Y., Small, J.P., Amori, M.E.S., and Kim, P.: Phys. Rev. Lett. **94**, 176803 (2005).

[42] Hwang, E.H., Adam, S., and Das Sarma, S.: Phys. Rev. Lett. **98**, 186806 (2007).

[43] Bolotin, K.I., Sikes, K.J., Jiang, Z., Fundenberg, G., Hone, J., Kim, P., and Stormer, H.L.: Sol. State Comm. **146**, 351 (2008).

[44] Du, X., Skachko, I., Barker, A., and Andrei, E.Y.: Nature Nano. **3**, 491 (2008).

[45] Dean, C.R., Young, A.F., Meric, I., Lee, C., Wang, L., Sorgenfrei, S., Watanabe, K., Taniguchi, T., Kim, P., Shepard, K.L., and Hone, J.: Nature Nano. **5**, 722 (2010).

[46] Cheianov, V.V., and Fal'ko, V.I.: Phys. Rev. B **74**, 041403 (2006).

[47] Sauter, F.: Zeitschrift für Physik A Hadrons and Nuclei **69**, 742 (1931).

[48] Lukose, V., Shankar, R., and Baskaran, G.: Phys. Rev. Lett. **98**, 116802 (2007).

[49] Huard, B., Stander, N., Sulpizio, J.A., and Goldhaber-Gordon, D.: Phys. Rev. B **78**, 121402 (2008).

[50] Shytov, A.V., Rudner, M.S., and Levitov, L.S.: Phys. Rev. Lett. **101**, 156804 (2008).

Philip Kim
Department of Physics
Harvard University
LISE 410
11 Oxford Street
Cambridge, MA 02138, USA
e-mail: `pkim@physics.harvard.edu`

Dirac Matter, 25–53

Dirac Fermions in Condensed Matter and Beyond

Mark Goerbig and Gilles Montambaux

Abstract. This review aims at a theoretical discussion of Dirac points in two-dimensional systems. Whereas Dirac points and Dirac fermions are prominent low-energy electrons in graphene (two-dimensional graphite), research on Dirac fermions in low-energy physics has spread beyond condensed matter to artificial graphene systems. In these alternative systems, a large versatility in the manipulation of the relevant band parameters can be achieved. This allows for a systematic study of the motion and different possible fusions of Dirac points, which are beyond the physical limits of graphene. We introduce the basic properties of Dirac fermions and the motion of Dirac points here and aim at a topological classification of these motions. The theoretical concepts are illustrated in particular model systems.

1. Introduction

During the last decade, graphene research has triggered a tremendous interest in the physics of two-dimensional (2D) Dirac fermions in condensed-matter physics [1]. Indeed, in undoped graphene the valence band touches the conduction band at the Fermi level isotropically in a linear manner. The low-energy band structure as well as the form of the underlying Hamiltonian are reminiscent of those for massless fermions, usually studied in high-energy physics, with two relevant differences. First, graphene electrons are constrained to move in two spatial dimensions, whereas the framework of relativistic quantum mechanics was established to describe fermions in three spatial dimensions. And second, the characteristic velocity that appears in condensed-matter physics is not the speed of light but the Fermi velocity, which is roughly two orders of magnitude smaller than the former. In addition to graphene, Dirac fermions have now been identified in various other systems, both in condensed matter physics, such as at the surfaces of three-dimensional topological insulators [2] or in quasi-2D organic materials [3], or in specially designed systems ("artificial graphenes"), such as, e.g., cold atoms in optical lattices [4], molecular crystals [5] or microwave crystals [6]. Even if they are probably less promising for technological applications than graphene, these artificial graphenes have the advantage that the relevant parameters can be varied more easily. This versatility is the main motivation of this, mainly theoretical, review on Dirac fermions in 2D systems, where we discuss the different manners of moving Dirac points in reciprocal space and where we aim at a classification of the different types of Dirac-point merging. Whereas we illustrate the theoretical concepts and this classification in several systems realized experimentally, we do not aim at a complete account of artificial graphenes and physical systems investigated in this framework.

The review is organized as follows. In Section 2, we discuss the general framework of two-band Hamiltonians that may display Dirac points and investigate the role of discrete symmetries, such as time-reversal and inversion symmetry. These considerations are the basis for an analysis of the underlying (two-component) tight-binding models that describe Dirac fermions (Section 3). In this section, we discuss both the topological properties of the spinorial wave functions in the vicinity of Dirac points and the specific case of graphene. Section 4 is devoted to the behaviour of Dirac fermions in a strong magnetic field and the relation between the topological winding properties of the wave functions and protected zero-energy levels. In Section 5, we discuss the motion and merging of Dirac points related by time-reversal symmetry and experimental implementations in cold atoms and microwave crystals (Section 6). We terminate this review with a more general discussion of how to obtain several pairs of Dirac points in tight-binding models and a second class of Dirac-point merging that is topologically different from that of Dirac points related by time-reversal symmetry (Section 7).

2. Emergence of Dirac fermions in a generic two-band model

In lattice models, Dirac fermions emerge at isolated points in the first Brillouin zone (BZ), where an upper band touches the lower one. The physically most interesting situation arises when the Fermi level precisely resides in these contact points, as for example in undoped graphene [1]. On quite general grounds, a two-band model that could reveal Dirac points may be expressed in terms of the band Hamiltonian

$$\mathcal{H}_{\vec{k}} = \sum_{\mu=0}^{3} f_{\vec{k}}^{\mu} \sigma^{\mu} = \begin{pmatrix} f_{\vec{k}}^0 + f_{\vec{k}}^z & f_{\vec{k}}^x - i f_{\vec{k}}^y \\ f_{\vec{k}}^x + i f_{\vec{k}}^y & f_{\vec{k}}^0 - f_{\vec{k}}^z \end{pmatrix} \tag{1}$$

in reciprocal space, where σ^0 is the 2×2 one matrix and

$$\sigma^x = \begin{pmatrix} 0 & 1 \\ 1 & 0 \end{pmatrix}, \qquad \sigma^y = \begin{pmatrix} 0 & -i \\ i & 0 \end{pmatrix}, \qquad \sigma^z = \begin{pmatrix} 1 & 0 \\ 0 & -1 \end{pmatrix}$$

are Pauli matrices. Because the band Hamiltonian must be a Hermitian matrix, the functions $f_{\vec{k}}^{\mu}$ are real functions of the 2D wave vector $\vec{k} = (k_x, k_y)$ that reflect furthermore the periodicity of the underlying lattice. The two bands are easily obtained from a diagonalization of Hamiltonian (1) that yields

$$\tilde{\epsilon}_\lambda(\vec{k}) = f_{\vec{k}}^0 + \lambda\sqrt{(f_{\vec{k}}^x)^2 + (f_{\vec{k}}^y)^2 + (f_{\vec{k}}^z)^2}, \tag{2}$$

where $\lambda = \pm 1$ is the band index. One notices that the function $f_{\vec{k}}^0$ is only an offset in energy, and we define for convenience the band energy $\epsilon_{\lambda,\vec{k}} = \tilde{\epsilon}_{\lambda,\vec{k}} - f_{\vec{k}}^0$ – since the function $f_{\vec{k}}^0$ in the Hamiltonian goes along with the one matrix, it does not affect the (spinorial) eigenstates obtained from a diagonalization of the Hamiltonian. One clearly sees from the generic expression (2) for the two bands, that band contact points require the annihilation of all three functions $f_{\vec{k}}^x$, $f_{\vec{k}}^y$ and $f_{\vec{k}}^z$ at isolated values $\vec{k}_D$ of the wave vector. In a 2D space, one is thus confronted with a system of three equations,

$$f_{\vec{k}_D}^x = 0, \qquad f_{\vec{k}_D}^y = 0, \qquad f_{\vec{k}_D}^z = 0,$$

to determine two values, i.e., the components of the wave vector $\vec{k}_D = (k_D^x, k_D^y)$. In contrast to three spatial dimensions, where the equations would determine three parameters and where one would then obtain three-dimensional Weyl fermions [7], stable band-contact points can only be obtained in 2D when one of the components is equal to zero over a larger interval of wave vectors. This situation arises when particular symmetries are imposed on the system that we discuss in the next paragraph.

One of the symmetries that protect stable band-contact points in Hamiltonian (1) is time-reversal symmetry, and it imposes

$$\mathcal{H}_{-\vec{k}}^* = \mathcal{H}_{\vec{k}} \qquad \rightarrow \qquad f_{-\vec{k}}^x = f_{\vec{k}}^x, \qquad f_{-\vec{k}}^y = -f_{\vec{k}}^y, \qquad f_{-\vec{k}}^z = f_{\vec{k}}^z. \tag{3}$$

Notice that in this argument, we have omitted the spin degree of freedom. Since we consider here a system with no spin-orbit coupling, each of the bands (2) is simply two-fold degenerate.[1] Another relevant symmetry is inversion symmetry. Consider that the diagonal elements of Hamiltonian (1) represent intra-sublattice (or intra-orbital) couplings. The first spinor component would then correspond to the weight on the A sublattice and the second one to that on the B sublattice. Inversion symmetry imposes that the Hamiltonian be invariant under the exchange of the two sublattices $A \leftrightarrow B$ while interchanging $\vec{k} \leftrightarrow -\vec{k}$, in which case one finds the conditions

$$f_{-\vec{k}}^x = f_{\vec{k}}^x, \qquad f_{-\vec{k}}^y = -f_{\vec{k}}^y, \qquad f_{-\vec{k}}^z = -f_{\vec{k}}^z, \tag{4}$$

for the periodic functions. Whereas the first two conditions are compatible with those (3) obtained for time-reversal symmetry, the presence of both symmetries (inversion and time-reversal) yields a function $f_{\vec{k}}^z$ that is both even and odd in the

---

[1] Otherwise, time-reversal symmetry would read $\mathcal{H}_{-\sigma,-\vec{k}}^* = \mathcal{H}_{\sigma,\vec{k}}$ where σ denotes the orientation of the electronic spin, and the band Hamiltonian is thus necessarily a 4×4 matrix. A detailed discussion of this case is beyond the scope of the present review.

wave vector, i.e., it must eventually vanish for all wave vectors, $f_{\vec{k}}^z = 0$. In the remainder of this review, we will therefore restrict the discussion to the Hamiltonian

$$\mathcal{H}_{\vec{k}} = \begin{pmatrix} 0 & f_{\vec{k}} \\ f_{\vec{k}}^* & 0 \end{pmatrix}, \tag{5}$$

which respects both symmetries and where we have defined the complex function $f_{\vec{k}} = f_{\vec{k}}^x - i f_{\vec{k}}^y$.

3. Dirac fermions in tight-binding models and fermion doubling

Before discussing the specific lattice model relevant for graphene, let us consider some general aspects of Hamiltonian (5) in the framework of general tight-binding parameters. Indeed the non-zero components of the band Hamiltonian reflect the periodicity of the Bravais lattice and may be written quite generally in the form

$$f_{\vec{k}} = \sum_{m,n} t_{mn} e^{-i\vec{k}\cdot\vec{R}_{mn}} , \tag{6}$$

where the hopping amplitudes t_{mn} are real, a consequence of the time-reversal symmetry in equation (3), and $\vec{R}_{mn} = m\vec{a}_1 + n\vec{a}_2$ are vectors of the underlying Bravais lattice. The Dirac points, which we coin **D** and **D′**, are solutions of the complex equation $f_{\vec{D}} = 0$ (see above). Since $f_{\vec{k}} = f_{-\vec{k}}^*$, the Dirac points, when they exist, necessarily come in by pairs, i.e., if $\vec{D}$ is a solution, so is $-\vec{D}$. Whereas this is a natural situation in lattice systems and gives rise to a $2N$-fold valley degeneracy (in the case of N pairs of Dirac points), it happens to be a problematic situation in high-energy physics, where Dirac fermions (of continuous systems) are sometimes simulated in lattice models and where fermions are thus doubled artificially [8]. The positions of the Dirac points can be anywhere in the BZ and move upon variation of the band parameters t_{mn}. Around the Dirac points $\pm\vec{D}$, the function $f_{\vec{k}}$ varies linearly. Writing $\vec{k} = \pm\vec{D} + \vec{q}$, we find, in a system of units with $\hbar = 1$ that we adopt henceforth,

$$f_{\pm\vec{D}+\vec{q}} = \vec{q} \cdot (\pm\vec{v}_1 - i\vec{v}_2), \tag{7}$$

where the velocities $\vec{v}_1$ and $\vec{v}_2$ are given by

$$\vec{v}_1 = \sum_{mn} t_{mn}\vec{R}_{mn}\sin\vec{D}\cdot\vec{R}_{mn} \quad \text{and} \quad \vec{v}_2 = \sum_{mn} t_{mn}\vec{R}_{mn}\cos\vec{D}\cdot\vec{R}_{mn}. \tag{8}$$

Furthermore, one generally has $\vec{v}_1 \neq \vec{v}_2$ so that the Dirac cones are not necessarily isotropic, and the low-energy Hamiltonian in the vicinity of the Dirac points $\xi\vec{D}$ is written as

$$\mathcal{H}_{\vec{q}}^{\xi} = \vec{q} \cdot (\xi\vec{v}_1\sigma^x + \vec{v}_2\sigma^y), \tag{9}$$

where ξ is the valley index ($\xi = +$ for **D** and $\xi = -$ for **D′**). Naturally, if there are other pairs of Dirac points, one obtains pairs of Hamiltonians of type (9) for each of them. The above analysis will further help us in the discussion of Dirac-point motion and merging presented in Section 5. In the vicinity of the Dirac points, the dispersion relation is then given by

$$\epsilon_\lambda(\vec{q}) = \lambda\sqrt{(\vec{v}_1 \cdot \vec{q})^2 + (\vec{v}_2 \cdot \vec{q})^2}. \tag{10}$$

Notice that the case $\vec{v}_1 \parallel \vec{v}_2$ is pathological in the sense that there would be no dispersion in the direction perpendicular to $\vec{v}_1$ and $\vec{v}_2$. This would be a quasi-1D limit that we exclude in the following discussions.[2]

3.1. Rotation to a simplified model and spinorial form of the wave functions

The Dirac Hamiltonian (9) may be further simplified and brought to the form

$$\mathcal{H}_{\vec{q}}^{\xi} = \xi v_x' q_x' \sigma^{x'} + v_y' q_y' \sigma^{y'} \tag{11}$$

with the help of a rotation of the coordinate space,

$$q_x = \cos\vartheta q_x' + \sin\vartheta q_y'$$
$$q_y = -\sin\vartheta q_x' + \cos\vartheta q_y',$$

accompanied by a rotation of the (sublattice)-pseudospin frame around the z-quantization axis

$$\sigma^x = \cos\theta\sigma^{x'} + \sin\theta\sigma^{y'}$$
$$\sigma^y = -\sin\theta\sigma^{x'} + \cos\theta\sigma^{y'},$$

and in terms of the novel velocities [3]

$$v_x'^2 = \frac{|\vec{v}_1|^2 + |\vec{v}_2|^2}{2} + \frac{1}{2}\sqrt{|\vec{v}_1|^4 + |\vec{v}_2|^4 + 2(\vec{v}_1 \cdot \vec{v}_2)^2 - 2(\vec{v}_1 \wedge \vec{v}_2)^2},$$
$$v_y'^2 = \frac{|\vec{v}_1|^2 + |\vec{v}_2|^2}{2} - \frac{1}{2}\sqrt{|\vec{v}_1|^4 + |\vec{v}_2|^4 + 2(\vec{v}_1 \cdot \vec{v}_2)^2 - 2(\vec{v}_1 \wedge \vec{v}_2)^2}.$$

Here, $\vec{v}_1 \wedge \vec{v}_2 = (\vec{v}_1 \times \vec{v}_2)_z$ is the z-component of the 3D vector product $\vec{v}_1 \times \vec{v}_2$. In the remainder of this section, we omit the primes at the velocities and wave vectors assuming that we are in the appropriate frame after transformation.

The eigenenergies of the low-energy model (11) are simply

$$\epsilon_\lambda(\vec{q}) = \lambda\sqrt{v_x^2 q_x^2 + v_y^2 q_y^2}, \tag{12}$$

with the corresponding eigenstates

$$\psi_{\xi,\lambda;\vec{q}} = \frac{1}{\sqrt{2}}\begin{pmatrix} 1 \\ \xi\lambda e^{-i\phi_{\vec{q}}} \end{pmatrix}, \tag{13}$$

where the relative phase between the two components depends on the orientation of the wave vector,

$$\tan\phi_{\vec{q}} = \frac{v_y q_y}{v_x q_x}. \tag{14}$$

---

[2]It would require an expansion of $f_{\vec{k}}$ beyond linear order around the Dirac points to obtain a dispersion in this direction.

3.2. Berry phases and winding numbers

The relative phase $\phi_{\vec{q}}$ derived above in equation (14) exhibits a particular topological structure encoded in the Berry phase [9]. Around each Dirac point, the circulation of $\phi_{\vec{q}}$ along a closed path is quantized; the quantity

$$w_{\xi,\lambda} = \frac{\xi\lambda\mathrm{sgn}(v_x v_y)}{2\pi} \oint_{C_i} \nabla_{\vec{q}}\phi_{\vec{q}} \cdot d\vec{q} \tag{15}$$

is an integer, the topological winding number associated with each Dirac point.[3] With the help of equation (14), one finds that the winding number associated with a constant-energy path at either positive ($\lambda = +1$) or negative ($\lambda = -1$) energy around the Dirac point ξ is simply

$$w_{\xi,\lambda} = \xi\lambda\mathrm{sgn}(v_x v_y). \tag{16}$$

An important observation is the fact that, because the two Dirac points $\mathbf{D}$ and $\mathbf{D}'$ are related by time-reversal symmetry, they have *opposite* winding numbers.

This argument may be generalized to situations with several pairs of Dirac points related by time-reversal symmetry. Consider N pairs of Dirac points (situated at the positions $\xi\vec{D}_i$ in the first Brillouin zone, $i = 1, \ldots, N$), each of which is described at low energy by a Hamiltonian of the type (9),

$$\mathcal{H}_{\vec{q}}^{i,\xi} = \vec{q} \cdot \left(\xi\vec{v}_1^i \sigma^x + \vec{v}_2^i \sigma^y\right). \tag{17}$$

Applying the same arguments around these novel points yields a winding number

$$w_{\xi,\lambda}^i = \xi\lambda\frac{\vec{v}_1 \wedge \vec{v}_2}{|\vec{v}_1 \wedge \vec{v}_2|}. \tag{18}$$

The winding number, which is a conserved quantity, may be interepreted as a topological charge. It is additive, and we will therefore use this concept extensively when discussing the different types of Dirac-point motion and merging in Section 5. Most saliently, it provides a simple and convenient manner of identifying the number of topologically protected Dirac points and zero-energy states, namely in a magnetic field, as we will show below.

3.3. Basic properties of electrons in graphene

Before discussing these zero-energy states in a magnetic field, let us use the above considerations to analyse the band structure of graphene [10], the probably best known instance of 2D Dirac points in condensed matter. Graphene consists of a one-atom thick layer of carbon atoms arranged in a honeycomb lattice (see left panel of Figure 1 for a sketch of the lattice structure). The underlying Bravais lattice is thus a triangular lattice, and the honeycomb structure is obtained with the help of a two-atom basis (sites A and B). The low-energy electronic properties

---

[3]This quantity is, modulo a factor of π, nothing other than the Berry phase accumulated on the path C_i around the Dirac point. However, it is more convenient to use the concept of (topological) winding numbers – whereas one is used, in basic quantum mechanics, to the fact that a phase 2π is identical to 0, the winding number is a physically relevant quantity (as we show below) that makes a clear distinction between $w_i = 0$ and 2.

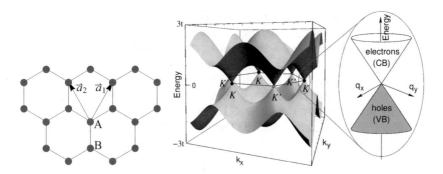

FIGURE 1. Left: Honeycomb lattice of graphene. The simplified tight-binding model takes into account hopping between nearest-neighbour sites, e.g., from A to B. In the case of undistorted graphene, the associated three hopping amplitudes are identical due to the point symmetry of the lattice. Right: Energy bands of graphene obtained from the tight-binding model and zoom around the Dirac point at K.

of graphene can be obtained to great accuracy within a simplified tight-binding model, where hopping only between nearest-neighbour p_z orbitals is taken into account, with a characteristic energy scale of $t \simeq 3$ eV. The band Hamiltonian (1) thus only has off-diagonal terms, and the function (6) simply reads [11]

$$f_{\vec{k}} = t \left(1 + e^{i\vec{k}\cdot\vec{a}_1} + e^{i\vec{k}\cdot\vec{a}_2} \right), \tag{19}$$

where $\vec{a}_1 = (\sqrt{3}a/2)(\vec{u}_x + \sqrt{3}\vec{u}_y)$ and $\vec{a}_2 = (\sqrt{3}a/2)(-\vec{u}_x + \sqrt{3}\vec{u}_y)$ are basis vectors that span the triangular Bravais lattice (see Figure 1). Here, $\vec{u}_x$ and $\vec{u}_y$ are the unit vectors in the x- and y- direction, respectively, and $a \simeq 0.14$ nm is the distance between nearest-neighbour carbon atoms. Notice that, because both sublattices A and B consist of carbon atoms, inversion symmetry is respected as well as time-reversal symmetry. Corrective tight-binding terms that are neglected in the present model therefore do not generate a non-zero term $f_{\vec{k}}^z$.[4] The band structure of graphene is depicted in Figure 1 (right pannel), and one notices the characteristic Dirac points at the corners of the first Brillouin zone,

$$\vec{K} = \frac{4\pi}{3\sqrt{3}a}\vec{u}_x \qquad \text{and} \qquad \vec{K}' = -\vec{K} = -\frac{4\pi}{3\sqrt{3}a}\vec{u}_x, \tag{20}$$

where the Fermi level resides in the absence of doping.[5] Notice furthermore that there are four additional contact points in the dispersion relation visible in Fig-

---

[4]Next-nearest-neighbour hopping breaks particle-hole symmetry by generating a term $f_{\vec{k}}^0$ but does not open a gap.

[5]Indeed, the electronically relevant p_z orbitals, which give rise to the electronic bands, are each occupied by a single electron such that the band structure is half-filled, and the Fermi level is thus situated at the Dirac points.

ure 1. However, these additional points are connected to $\vec{K}$ and $\vec{K}'$ by a reciprocal lattice vector so that they correspond to the same electronic state. The fact that the Dirac points coincide with the corners of the Brillouin zone in the case of graphene is a consequence of the crystal point symmetry – as we will show in the following sections, deviations from this symmetry place the Dirac points at less symmetric points of the BZ, but the Dirac points are nevertheless topologically protected by inversion and time-reversal symmetry.

Similarly to the general case, the low-energy Hamiltonian is readily obtained by expanding $f_{\vec{k}}$ as in equation (7) around the Dirac points,

$$f_{\pm\vec{D}+\vec{q}} = v_F(\xi q_x - iq_y), \tag{21}$$

where $v_F = 3at/2$ is the Fermi velocity. Compared to equation (7), the point symmetry of the undistorted graphene lattice provides us with isotropic Dirac points, $\vec{v}_1 = (v_F, 0)$ and $\vec{v}_2 = (0, v_F)$, and Hamiltonian (9) becomes

$$\mathcal{H}_{\vec{q}}^{\xi} = v_F(\xi q_x \sigma^x + q_y \sigma^y). \tag{22}$$

This Hamiltonian is precisely that of massless Dirac fermions of relativistic quantum mechanics in two spatial dimensions. The description of low-energy electrons in graphene within this framework has been extremely successful in identifying its original electronic properties. A full account on these properties is beyond the scope of the present review, and we refer the reader to existing reviews. Apart from the general ones [1, 12], there are reviews that concentrate on more specific aspects of graphene, such as [13] and [14] for electronic transport, [15] and [16] for experimental reviews, as well as [17] and [18] for interaction effects in graphene.

4. Dirac fermions in a magnetic field

Instead of providing an exhaustive account on physical phenomena of graphene electrons, we discuss here the consequences of the above-mentioned topological aspects in the presence of a magnetic field. Most saliently, the winding numbers allow us to identify the degeneracy of the zero-energy states (zero-energy Landau levels) and their topological stability.

4.1. Landau levels of Dirac fermions

The basic model Hamiltonian (11), transformed to the appropriate reference frame as described above, is amenable to an exact quantum-mechanical solution when taking into account a perpendicular magnetic field via the Peierls substitution,

$$\vec{q} \to \vec{\Pi} = \vec{p} + e\vec{A}(\vec{r}), \tag{23}$$

where $\vec{p}$ is the quantum-mechanical momentum (in the continuum limit), which is conjugate to a coarse-grained position operator $\vec{r}$. This Peierls substitution is valid as long as the magnetic field $B = |\nabla \times \vec{A}(\vec{r})|$ is sufficiently small, such that the associated magnetic length $l_B = 1/\sqrt{eB} \simeq 26\,\text{nm}/\sqrt{B[\text{T}]}$ is much larger than the lattice spacing, $l_B \gg a$. This condition is ususaly satisfied in condensed matter physics at accessible magnetic fields (up to 45 T for static magnetic fields and $B \lesssim 200$ T for pulsed fields in the semi-destructive regime). The quantum-mechanical commutation relations $[x, p_x] = [y, p_y] = i$ (and 0 for the crossed commutators) induce the commutation relations

$$[\Pi_x, \Pi_y] = -\frac{i}{l_B^2} \tag{24}$$

for the components of the gauge-invariant kinetic momementum. In terms of the associated ladder operators

$$\hat{a} = \frac{l_B}{\sqrt{2}}(\Pi_x - i\Pi_y) \quad \text{and} \quad \hat{a}^\dagger = \frac{l_B}{\sqrt{2}}(\Pi_x + i\Pi_y), \tag{25}$$

which satisfy $[\hat{a}, \hat{a}^\dagger] = 1$, the Peierls substitution can be simplified to

$$q \to \frac{\sqrt{2}}{l_B}\hat{a} \quad \text{and} \quad q^* \to \frac{\sqrt{2}}{l_B}\hat{a}^\dagger, \tag{26}$$

and the Hamiltonian (11) can eventually be written as

$$H_B^{\xi=+} = \sqrt{2}\frac{v_F}{l_B}\begin{pmatrix} 0 & \hat{a} \\ \hat{a}^\dagger & 0 \end{pmatrix} \quad \text{and} \quad H_B^{\xi=-} = -\sqrt{2}\frac{v_F}{l_B}\begin{pmatrix} 0 & \hat{a}^\dagger \\ \hat{a} & 0 \end{pmatrix}, \tag{27}$$

where the Fermi velocity v_F is an average quantity $v_F = \sqrt{v_x v_y}$.

The energy spectrum of the so-called Landau levels can be obtained directly with the help of the eigenstates of the number operator $\hat{n} = \hat{a}^\dagger \hat{a}$,

$$\epsilon_{\lambda,n} = \lambda\frac{v_F}{l_B}\sqrt{2n}, \tag{28}$$

and reveals the characteristic $\sqrt{Bn}$ scaling [19] observed in magneto-spectroscopic measurements [20] and scanning-tunneling spectroscopy [21]. Notice that the spectrum does not depend on the valley index ξ, and one obtains thus a two-fold valley degeneracy in addition to the usual spin degeneracy (unless the latter is lifted by a strong Zeeman effect). The corresponding eigenstates read

$$\psi_{\lambda,n\neq0}^{\xi=+} = \frac{1}{\sqrt{2}}\begin{pmatrix} |n-1,m\rangle \\ \lambda|n,m\rangle \end{pmatrix} \quad \text{and} \quad \psi_{\lambda,n\neq0}^{\xi=-} = \frac{1}{\sqrt{2}}\begin{pmatrix} |n,m\rangle \\ -\lambda|n-1,m\rangle \end{pmatrix}, \tag{29}$$

for the levels with $n \neq 0$. Here, the spinor components satisfy $\hat{n}|n,m\rangle = n|n,m\rangle$, and an additional quantum number m needs to be taken into account to complete the basis. We discuss the physical meaning of this quantum number below. The zero-energy Landau levels $n = 0$ need to be treated separately and reveal a very special structure,

$$\psi_{n=0}^{\xi=+} = \begin{pmatrix} 0 \\ |n=0,m\rangle \end{pmatrix} \quad \text{and} \quad \psi_{n=0}^{\xi=-} = \begin{pmatrix} |n=0,m\rangle \\ 0 \end{pmatrix}. \tag{30}$$

One notices that, at zero energy, the valley degree of freedom coincides with the sublattice index. At zero energy, the dynamical properties of the A sublattice (electrons in valley K') are thus completely decoupled from that on the B sublattice (electrons in valley K). This gives rise to the particular series of quantum Hall effects at filling factors

$$\nu = \frac{n_{el}}{n_B} = 2\pi l_B^2 n_{el} = \pm 2(2n+1), \qquad (31)$$

where n_{el} is the 2D electronic density and $n_B = 1/2\pi l_B^2 = eB/h$ the density of flux quanta threading the graphene sheet. The effect manifests itself by plateaus in the transverse (Hall) resistance, at magnetic fields corresponding to these filling factors, accompanied by zeros in the longitudinal resistance. The observation of quantum Hall states at the filling factors (31) in 2005 [22, 23] was interpreted as a direct proof of the presence of pseudo-relativistic carriers in graphene described in terms of massless Dirac fermions (22).

4.2. Degeneracy

We have already alluded to the presence of a second quantum number m in the full description of the quantum-mechanical states (29) and (30). Since the Landau-level spectrum (28) does not depend on this quantum number, its presence yields a degeneracy of the Landau levels that we discuss in more detail here. Indeed, the kinetic momentum $\vec{\Pi}$ introduced in the Peierls substitution (23) may be related to the cyclotron variable $\vec{\eta}$ with the components

$$\eta_x = l_B^2 \Pi_y \qquad \text{and} \qquad \eta_y = -l_B^2 \Pi_x, \qquad (32)$$

that satisfy in turn the commutation relation $[\eta_x, \eta_y] = -il_B^2$. In classical mechanics, this cyclotron variable describes precisely the cyclotron motion of a charged particle in a uniform magnetic field. In addition, one knows from classical mechanics that the particle's energy does not depend on the position of the centre of the cyclotron motion, which is thus a constant of motion. This gauge-invariant centre of the cyclotron motion,

$$\vec{R} = (X, Y) = \vec{r} - \vec{\eta}, \qquad (33)$$

which is also called *guiding centre*, remains a constant of motion in the quantum-mechanical description, i.e., its components commute with the Hamiltonian

$$[X, H_B^\xi] = [Y, H_B^\xi] = 0.$$

Furthermore, one can show from the decomposition (33) that the components of the guiding centre commute with those of the cyclotron variable

$$[\eta_x, X] = [\eta_y, X] = [\eta_x, Y] = [\eta_y, Y] = 0, \qquad (34)$$

whereas the guiding-centre coordinates do not commute among each other,

$$[X, Y] = il_B^2. \qquad (35)$$

This allows for the introduction of a second set of ladder operators

$$\hat{b} = \frac{1}{\sqrt{2}l_B}(X + iY) \qquad \text{and} \qquad \hat{b}^\dagger = \frac{1}{\sqrt{2}l_B}(X - iY), \qquad (36)$$

with $[\hat{b}, \hat{b}^\dagger] = 1$, similarly to those (25) introduced in the description of the kinetic momentum. The second quantum number m is thus simply the eigenvalue of $\hat{b}^\dagger \hat{b}$, and describes, as mentioned above, the orbital degeneracy of the Landau levels. Instead of deriving explicitly this degeneracy,[6] one may invoke an argument via the Heisenberg uncertainty relation associated with the commutation relation (35),

$$\Delta X \Delta Y \gtrsim 2\pi l_B^2 = \sigma. \qquad (37)$$

This means that each quantum-mechanical state $|n, m\rangle$ occupies a minimal surface $\sim \sigma$, and the degeneracy of each Landau level may thus be quantified by dividing the full area Σ by this minimal surface,

$$N_B = \frac{\Sigma}{\sigma} = n_B \Sigma, \qquad (38)$$

in terms of the flux density $n_B = 1/2\pi l_B^2 = eB/h$, which we have already encountered in the previous paragraphs. The filling factor (31) can thus be interpreted as the number of Landau levels that are completely filled, while not taking into account its internal degeneracy due to the spin and valley degrees of freedom. The latter four-fold degeneracy indicates that there are eventually $4N_B$ states per Landau level and the quantum-Hall plateaus thus occur at multiples of four of the filling factor, as suggested by equation (31). Notice finally that, in principle, the Heisenberg uncertainty relation (37) is an inequality and that σ would then just be a lower bound of the surface occupied by a quantum state. However, the above-mentioned calculation in a special geometry and a special gauge [24] indicates that the minimal surface is σ and that the degeneracy is indeed given by equation (38). As a qualitative explanation of this fact, one may invoke the harmonic-oscillator structure of the quantum-mechanical system – the associated wave functions are therefore Gaussians for which the Heisenberg uncertainty relation becomes an equality.

Notice finally that the arguments in this subsection rely only on the algebraic structure of 2D electrons in a magnetic field and not on the precise form of the Hamiltonian. As long as the latter can be expressed solely in terms of the kinetic-momentum operator (Π_x, Π_y), each of its (perhaps unspecified) energy levels is N_B-fold degenerate, apart from internal degrees of freedom, such as the spin, or a "topological" degeneracy that we will discuss in the following subsection. This particular feature is simply due to the existence of a second set of operators, X and Y, that commute with Π_x and Π_y but that do not commute among each other [equation (35)].

4.3. Semi-classical quantization rule

In the previous subsections, we have discussed the Landau-level spectrum of mass-less Dirac fermions in monolayer graphene and shown that these levels are highly degenerate as a consequence of the existence of a set of operators, X and Y, that commute with the Hamiltonian. However, many electronic systems do not have a simple low-energy description in terms of the 2D Dirac Hamiltonian (9) (or a

---

[6]Whereas the general proof is rather involved, the degeneracy can be obtained when analysing the wave functions in a particular gauge [24].

simple Schrödinger equation) that allows for an exact quantum-mechanical solution in the presence of a magnetic field. In this case, one needs to appeal to other methods, such as the semi-classical quantization rule [25, 26]

$$\mathcal{A}_\mathcal{C}(\epsilon_n) = \frac{2\pi}{l_B^2}\left(n + \frac{1}{2} - \frac{|w_\mathcal{C}|}{2}\right), \tag{39}$$

where $\mathcal{A}_\mathcal{C}(\epsilon_n)$ is the area in reciprocal space delimited by a closed contour $\mathcal{C}$, associated with energy ϵ_n. Furthermore, $w_\mathcal{C}$ represents the total winding of the relative phase $\phi_{\vec{k}}$ between the spinor components along this closed contour. In the case of massless Dirac fermions, this phase was given by equation (14), whereas in the general two-band model (11) it reads

$$\tan\phi_{\vec{k}} = \frac{\mathrm{Im}f_{\vec{k}}}{\mathrm{Re}f_{\vec{k}}}, \tag{40}$$

in terms of the real and the imaginary parts of the complex function $f_{\vec{k}}$.

One can easily show that the calculation of the reciprocal-space area $\mathcal{A}_\mathcal{C} = 2\pi\int_0^{k_n} dk\, k = 2\pi\int_0^{\epsilon_n} d\epsilon\, k(\epsilon)(dk/d\epsilon)$ yields the graphene Landau-level spectrum (28) for isotropic Dirac points, with $\epsilon = v_F k$, for which we have already calculated the winding number, $w = \pm 1$. Beyond the calculation of the Landau-level spectrum, the semi-classical analysis (39) is also extremely convenient in the classification of the different types of Dirac point motion and merging discussed in the following section. Indeed, the winding numbers are topological charges and thus conserved quantities in the different merging scenarios. The merging of two Dirac points with opposite winding numbers therefore gives rise to a total winding of $w = 0$, that is zero topological charge, and that of Dirac points with like winding yield a total topological charge of $|w| = 2$. We will discuss both types of merging transitions extensively in the following sections and finish this paragraph with another consequence of the topological charge. Indeed, it classifies the number of topologically protected zero-energy levels, which is

$$w_p = \left|\sum_i w_i\right|, \tag{41}$$

in the case of $i = 1, \ldots, 2N$ Dirac points in the system. As we have already mentioned, if the system is time-reversal symmetric (in the absence of a magnetic field), this number is necessarily zero because Dirac points emerge in pairs of opposite charge. However, if the Dirac points are sufficiently isolated at low magnetic fields, there are

$$w_t = \sum_i |w_i| \tag{42}$$

(i.e., $w_t = 2N$) zero-energy levels associated with the N pairs of Dirac Hamiltonians of the type (27) that describe the low-energy electronic excitations of the system.

5. Motion and merging of time-reversal-symmetric Dirac points

In the previous sections, we have seen that Dirac points appear as topological objects characterized by a charge. This charge describes the winding of the wave function in reciprocal space when turning around a Dirac point. In this section, we describe how a pair of such Dirac points with opposite charges related by time-reversal symmetry can move in reciprocal space or even be annihilated. As mentioned above for the case of the honeycomb lattice relevant in the description of a graphene crystal, the two Dirac points are precisely located at two opposite corners of the Brillouin zone (Figure 2). The position of the Dirac points $\vec{D} =$

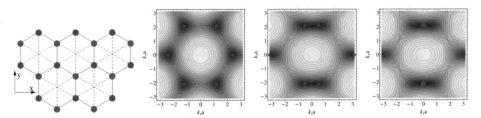

FIGURE 2. Left: honeycomb crystal. The vertical thick lines represent the modified hopping integral t'. The dashed lines indicate the third-nearest-neighbour coupling t_3 discussed in Section 7.1. Right: variation of the isoenergy lines for three values of $t'/t = 1, 1.8, 2$ ($t_3 = 0$). The red dots indicate the position of the two inequivalent Dirac points which merge when $t' = 2t$.

$\pm(\vec{a}_1^* - \vec{a}_2^*)/3$, where $\vec{a}_1^*$ and $\vec{a}_2^*$ are elementary reciprocal lattice vectors, at these high-symmetry points is, however, a rather exceptional situation. Consider for example the "brickwall" lattice depicted on Figure 3-a. It has the same couplings between sites as in graphene, but due to the square symmetry, the Brillouin zone is a square. The Dirac points are now located *inside* the first Brillouin zone (BZ) (Figure 3-b). Although this brickwall crystal may appear impossible to realize in

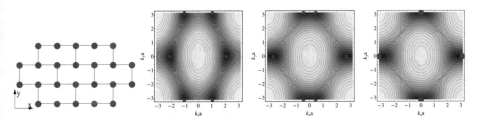

FIGURE 3. Left: brickwall crystal. The vertical thick lines indicate the modified hopping integral t'. Right: variation of the isoenergy lines for three values of $t'/t = 1, 1.8, 2$.

condensed matter, it has recently been realized for a crystal of cold atoms in an optical lattice [4], as we will discuss later in this review.

Moreover, upon variation of the band parameters, the two Dirac points may approach each other and merge into a single point $\mathbf{D}_0$. This happens when $\vec{D} = -\vec{D}$ modulo a reciprocal lattice vector $\vec{G}$. Therefore, the location of this merging point is simply $\vec{D}_0 = \vec{G}/2$. There are then *four* possible inequivalent points whose coordinates are $\vec{D}_0 = (p\vec{a}_1^* + q\vec{a}_2^*)/2$, with $(p, q) = (0, 0), (1, 0), (0, 1), (1, 1)$. These are precisely the time-reversal-invariant momenta (TRIM) of a 2D BZ. The condition for the existence of Dirac points, $f_{\vec{D}_0} = \sum_{mn}(-1)^{\beta_{mn}} t_{mn} = 0$, where $\beta_{mn} = pm + qn$, defines a manifold in the space of band parameters. This manifold separates a semi-metallic phase with two Dirac cones from a band insulator. One notices that the merging of Dirac points may even occur at the Γ point, under the condition that the hopping parameters do not have the same sign. This sign change may, e.g., be achieved in shaken optical lattices [27].

Remarkably, at the merging point, the velocity $\vec{v}_1$ vanishes, since

$$\sin(\vec{G} \cdot \vec{R}_{mn}/2) = 0,$$

so that the dispersion becomes massive along this direction, that we define as the x direction. This is a direct consequence of the form of the low-energy Hamiltonian in the vicinity of a TRIM. In order to respect time-reversal symmetry, it must satisfy, in terms of the continuum wave vector $\vec{q} = \vec{k} - \vec{D}_0$, the same structure (3) [and (4) in the case of an inversion-symmetric system] as the full band Hamiltonian (5) around the central Γ point. Therefore, to lowest order, the Hamiltonian may be expanded as

$$\mathcal{H}_0(\vec{q}) = \frac{q_x^2}{2m^*} \sigma^x + c_y q_y \sigma^y, \tag{43}$$

where the velocity c_y and the effective mass m^* may be related to the microsopic parameters (6) of the original Hamiltonian [28]. The terms of order q_y^2 and $q_x q_y$ are neglected at low energy [28, 29]. Most saliently, the corresponding energy spectrum

$$\epsilon = \pm \left[c_y^2 q_y^2 + \left(\frac{q_x^2}{2m^*} \right)^2 \right]^{1/2} \tag{44}$$

is linear in one direction and quadratic in the other, and the hybrid band-contact point has also been called a *semi-Dirac* point [30].

The merging of the Dirac points in $\mathbf{D}_0$ marks the transition between a semi-metallic phase and an insulating phase, and it can be analysed topologically in terms of winding numbers – below the transition, the semi-metallic phase is characterized by two Dirac points of opposite charges that are annihilated at the transition. The resulting zero topological charge then allows for the opening of a gap in the spectrum and a transition to an insulating phase. In order to describe the transition more quantitatively, we introduce the gap parameter

$$\Delta_* = f_{\vec{D}_0} = \sum_{mn} (-1)^{\beta_{mn}} t_{mn}, \tag{45}$$

which changes its sign at the transition. In the vicinity of the transition, the Hamiltonian has the universal form

$$\mathcal{H}_{+-}(\vec{q}) = \begin{pmatrix} 0 & \Delta_* + \frac{q_x^2}{2m^*} - ic_y q_y \\ \Delta_* + \frac{q_x^2}{2m^*} + ic_y q_y & 0 \end{pmatrix}, \tag{46}$$

with the spectrum

$$\epsilon = \pm\sqrt{\left(\Delta_* + \frac{q_x^2}{2m^*}\right)^2 + q_y^2 c_y^2}. \tag{47}$$

This *Universal Hamiltonian*, which describes the merging of two Dirac points with opposite charges [28], is universal in the sense that its structure is general, independent of the microscopic parameters. It has a remarkable structure and describes properly the vicinity of the topological transition, as shown on Figure 4-a. When Δ_* is negative (we choose $m^* > 0$ without loss of generality), the spectrum exhibits the two Dirac cones at a distance $2q_D = 2\sqrt{-2m^*\Delta_*}$ and a saddle point at $\vec{D}_0$ (the energy of the saddle point being $\pm|\Delta_*|$). Increasing Δ_* from negative to positive values, the saddle point shifts to lower energies and eventually disappears into the hybrid semi-Dirac point at the transition ($\Delta_* = 0$), before a gap of size $2\Delta_* > 0$ occurs in the spectrum.

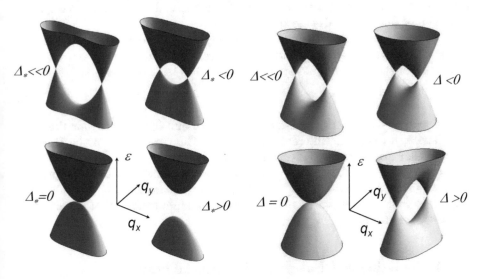

FIGURE 4. Universal scenario for the merging of two Dirac points a) with opposite winding numbers; b) with the same winding number. A gap may open in the first case, but not in the second one. The first case describes the merging of two Dirac points in a strained honeycomb lattice. The second case describes the evolution of the spectrum in twisted bilayer graphene, neglecting the trigonal warping (see Section 7.2).

Therefore, the Hamiltonian describes the continuous evolution from isolated Dirac points like in graphene to massive particles in the gapped phase. In particular, the Landau-level spectrum in a magnetic field evolves continuously from the well-known $\sqrt{nB}$ spectrum as in graphene to a massive particle spectrum $(n + 1/2)B$ above the merging transition, as one may see from the semi-classical quantization rule (39). Whereas below the transition there exist, at sufficiently low magnetic fields, closed orbits encircling just one of the Dirac points (with a winding number $|w_i| = 1$) and one therefore obtains a doubly degenerate zero-energy level, the situation is drastically different above the transition. Indeed, all possible orbits then have a winding number $w = 0$ that yields the $1/2$ offset in the Landau-level spectrum, as in the case of conventional Schrödinger fermions. Directly at the transition, the level spectrum shows an unusual behaviour $[(n + 1/2)B]^{2/3}$. Again the $1/2$ offset is due to the absence of closed orbits encircling singular points with $w \neq 0$ at the transition. Notice, however, that the topological transition in the presence of a magnetic field is not abrupt. Also below the transition, where the zero-field spectrum reveals two Dirac points, the two Dirac points are coupled by the magnetic field – indeed, the closed orbits necessarily enclose surfaces of size $1/l_B^2 \propto B$ in reciprocal space, due to the non-commutativity of the kinetic-moment operators (23), and, at sufficiently large magnetic fields, Dirac points separated by small wave vectors are no longer resolved. This lifts the original two-fold degeneracy of the zero-energy level $n = 0$ in an exponential manner. The continuous evolution of this level is discussed in Ref. [29].

As a simple example, the motion and merging of Dirac points may be realized in the above honeycomb and brickwall lattices with nearest-neighbour coupling, where one of the coupling parameters named t' has been increased [11]

$$f_{\vec{k}} = t(\beta + e^{i\vec{k}\cdot\vec{a}_1} + e^{i\vec{k}\cdot\vec{a}_2}), \qquad \text{with } \beta = t'/t, \tag{48}$$

with $\vec{a}_i = (\pm\frac{\sqrt{3}}{2}a, \frac{3}{2}a)$ for the honeycomb lattice and $\vec{a}_i = (\pm a, a)$ for the brickwall lattice, a being the interatomic distance. The parameters of the Universal Hamiltonian are then respectively $\Delta_* = t' - 2t, m^* = 2/(3ta^2), c_y = 3ta$ and $\Delta_* = t' - 2t, m^* = 1/(2ta^2), c_y = ta$. The merging scenario initially proposed in strained graphene [28, 29, 31, 32] turned out to be unreachable [33], but it has been observed in various systems, now called "artificial graphenes", that we discuss in the following section.

6. Manipulation of Dirac points in artificial graphenes

The intensive study of Dirac fermions in graphene has motivated the search for different systems sharing similar properties with graphene, in particular to exhibit phenomena which could not be observed in graphene. The flexibility of such systems may allow for the realization of properties unreachable in graphene, like the predicted topological transition or the manipulation of edge states. Examples for these artificial graphenes comprise photonic or microwave crystals [6, 34, 35, 36],

molecular crystals [5], ultracold atoms in optical lattices [4], polaritons propagating
in a honeycomb lattice of coupled micropillars etched in a planar semiconductor
microcavity [37], or the quasi-2D organic salt α-(BEDT-TTF)$_2$I$_3$ under pressure
[3]. We do not elaborate further on this now long list of different physical systems
(for a review, see Ref. [38]), but here we restrict the discussion to only two physical
systems where the manipulation and merging of Dirac points has been explicitly
observed and studied.

6.1. A lattice of cold atoms

Ultracold atoms trapped in an optical lattice offer beautiful realizations of condensed matter situations. It has recently been made possible to create a periodic
potential with the help of standing optical (laser) waves that trap cooled atoms
via a dipolar interaction. These trapped atoms can be described to great accuracy
within a tight-binding model simulating very closely the physics of graphene. The
lattice is indeed very close to the brickwall lattice depicted on Figure 3. By varying
appropriately the intensities of the laser fields, it is possible to realize exactly the
merging scenario described by the Universal Hamiltonian (46) [28, 39]. In order
to probe the spectrum, the position of the Dirac points, their motion and their
merging, a low-energy cloud of fermionic atoms is submitted to a constant force
F (Figure 5-a, b), so that its motion is uniform in reciprocal space $\hbar(d\vec{k}/dt) = \vec{F}$,
and exhibits Bloch oscillations [4]. In the vicinity of a Dirac point, there is a finite
probability for the atoms to tunnel into the upper band. This probability depends
on the applied force and on the gap separating the two bands. For a single crossing, it is given by Landau–Zener (LZ) theory [40]. By measuring the proportion
of atoms having tunneled into the upper band after one Bloch oscillation, it is
in principle possible to reveal the energy spectrum. Since the spectrum exhibits
a *pair* of Dirac points, it is important to separate two cases, as it has been done
experimentally.[7]

– *Single LZ tunneling.* In this case, the force is applied along the y-direction
perpendicular to the merging line and the cloud of atoms "hits" the two Dirac
points in parallel (Figure 5-a,c). An atom, initially in a state with finite q_x, per-
forms a Bloch oscillation along a line of constant q_x and may tunnel into the upper
band with a probability given by

$$P_Z^y = e^{-\pi \frac{(\text{gap}/2)^2}{c_y F}} = e^{-\pi \frac{(\frac{q_x^2}{2m^*} + \Delta_*)^2}{c_y F}}, \qquad (49)$$

where we have written the gap in terms of the parameters of the Universal Hamiltonian (46). In the gapped (G) phase, above the merging transition ($\Delta_* > 0$),
the tunneling probability is vanishingly small. In the opposite case, deep in the
Dirac (D) phase ($\Delta_* < 0$), when the distance $2q_D = 2\sqrt{-2m^*\Delta_*}$ between the
Dirac points is larger than the size of the cloud, the probability to tunnel into the

---

[7] Here we define direction x and y consistent with the rest of the paper. They are interchanged
compared to Refs. [4, 39].

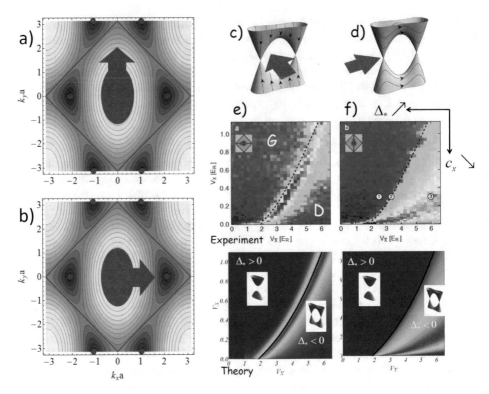

FIGURE 5. a, b) Schematic picture of the low energy fermionic cloud in reciprocal space. In case a, c), the force is applied in a direction perpendicular to the merging direction so that the fermionic cloud "hits" the two Dirac points in parallel. In case b, d), the force is applied along the merging direction so that the cloud encounters the two Dirac points in series, leading to a double Landau–Zener tunneling. e, f) Experimentally measured and calculated probability to tunnel to the upper band (blue indicates low and red high probabilities) as a function of laser-beam amplitude. In the case of double LZ tunneling, the probability is maximal inside the Dirac phase. The experimental data are adapted from Ref. [4] and the theoretical graphs from Ref. [39]

upper band is also small. The tunnel probability is actually large near the merging transition. Figure 5-e represents the tunneling probability as a function of two parameters of the lattice potential $(V_X, V_{\overline{X}})$ that we do not explicit here [4]. This result has been obtained by relating these parameters to the parameters of the Universal Hamiltonian and by using equation (49) [39]. The agreement with the experiment is excellent without any adjustable parameter (Figs. 5-e). To account quantitatively for the experimental result, one has to average the probability (49) over a finite range of q_x, due to the finite size of the fermionic cloud in reciprocal space (Figure 5-a). By doing so, one sees from (49) that the probability is maximal for a negative value of the driving parameter $\Delta_* = -\langle q_x^2 \rangle / 2m^*$ where $\langle \cdots \rangle$ is an appropriate average. This explains why the intensity is maximal *inside* the D phase, as seen in Figs. 5-e.

   – *Double LZ tunneling.* In this case the force F is applied along the merging x-direction and the cloud of atoms "hits" the two Dirac points in series (Figure 5-b,d). This situation is more involved since each atom may undergo two LZ transitions in a row. Each LZ transition is described by the tunnel probability

$$P_Z^x = e^{-\pi \dfrac{(\text{gap}/2)^2}{c_x F}} = e^{-\pi \dfrac{c_y^2 q_y^2}{c_x F}} = e^{-\pi \dfrac{c_y^2 q_y^2}{F\sqrt{2|\Delta_*|/m^*}}}. \tag{50}$$

Assuming that the two tunneling events are incoherent, the interband transition probability resulting from the two events in series is

$$P_t^x = 2P_Z^x(1 - P_Z^x). \tag{51}$$

   In the G phase, the tunneling probability is again vanishingly small. In the D phase, when $q_y = 0$, the single LZ probability is maximal ($P_Z^x = 1$), but the tunneling probability after two events vanishes. For an initial cloud of finite size q_y, the transferred fraction is an average $\langle P_t^x \rangle$ taken on the finite width of the cloud. The interband transition probability [equation (51)] is a non-monotonic function of the LZ probability P_Z^x, and it is maximal when $P_Z^x = 1/2$. This explains why the maximum of the tunnel probability is located well *inside* the D phase (red region in Figs. 5-f). Varying the averaging order, this happens when $\langle P_t^x \rangle \simeq 1/2$, that is for a finite value Δ_* given by $m^* c_y F \ln 2/(2\pi \langle q_x^2 \rangle)$, that is well inside the Dirac phase, as shown in Figs. 5-f.

   A particularly interesting effect related to the double LZ tunneling is the possible existence of interference effects between the two LZ events. Assuming the phase coherence is preserved, instead of the probability given by equation (51), one expects a resulting probability of the form

$$P_t^x = 4P_Z^x(1 - P_Z^x)\cos^2(\varphi/2 + \varphi_d), \tag{52}$$

where φ_d is a phase delay, named Stokes phase, attached to each tunneling event. Furthermore, $\varphi = \varphi_{dyn} + \varphi_g$ is a phase which has two contributions, a dynamical phase φ_{dyn} acquired between the two tunneling events and basically related to the energy difference between the two energy paths, and a geometric phase φ_g. Whereas the dynamical phase carries information about the spectrum, the geometric phase carries information about the structure of the wave functions [41]. It is now an experimental challenge to access directly this interference pattern and to probe the different contributions to the dephasing.

6.2. Propagation of microwaves

The rich physical properties associated with the propagation of electrons in a honeycomb lattice may also be revealed in the propagation of *any* wave in this lattice. Therefore electrons may be replaced by other waves such as light, microwaves, or other elementary excitations like polaritons. This may allow for a much more flexible realization of the same physics, but implying different length scales. As an example, we consider here a microwave that is confined between two metallic plates realizing a 2D situation and that propagates through an ensemble of dielectric cylindric dots of centimeter size (Figure 6-a). The frequency is chosen

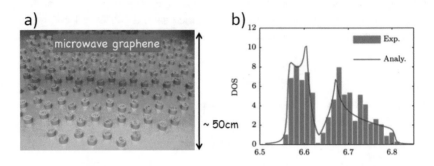

FIGURE 6. a) Honeycomb lattice of 288 dielectric cylinders. b) Experimental DOS well fitted by a tight-binding model with second and third-nearest-neighbour couplings.

such that the propagation is resonant inside a dot and evanescent outside the dots [35, 6]. Therefore the dots are weakly coupled through evanescent waves and the wave propagation between the dots is very well described within a tight-binding model [6, 42]. The signal is emitted and measured by an antenna which gives direct access to the local density of states (DOS). The measured DOS is plotted in Figure 6-b, and it can be described within a tight-binding model where second and third-nearest-neighbour couplings are not small (they depend on the distance between the dots and typically $t_2/t \simeq 0.09, t_3/t \simeq 0.07$). A uniaxial strain is easily realized in this setup, so that one of the couplings t' may be modified and typically the ratio $\beta = t'/t$ has been varied between 0.4 and 3.5. By doing so, the merging transition has been reached for a critical value $\beta_{cr} \simeq 1.8$ which corresponds very well the theoretically expected value taking into account the higher-order nearest-neighbour couplings $\beta_{cr} = 2 - 3t_3/t$, since from equation (45), we have here $\Delta_* = t' - 2t + 3t_3$.

The great advantage of this setup is its flexibility. It is quite easy to manipulate the "atoms" and to measure the local DOS. This flexibility has been used to modify at will the structure of the edges and to investigate the existence of edge states whose importance is well known in graphene. Indeed, zigzag edges support edge states while armchair edges do not. It has been predicted however that edge states may exist even in the armchair case, in the presence of uniaxial anisotropy [43]. The existence of edge states is clearly revealed by DOS measurement as seen on Figure 6-d when the anisotropy increases. It has been found that (i) edge states appear only along the edges that are not parallel to the anisotropy

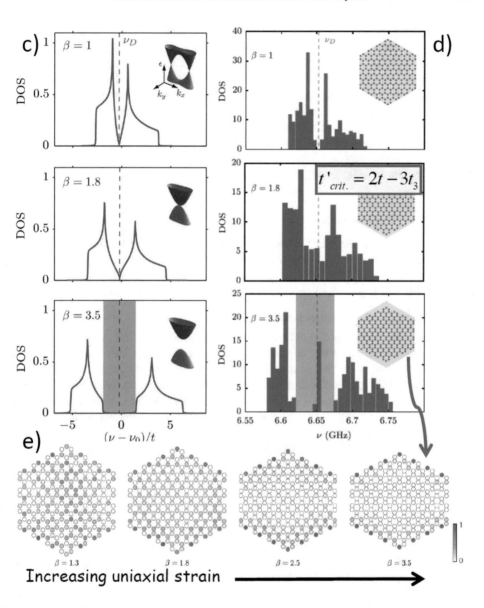

FIGURE 6. c) Expected evolution of the DOS with anisotropy of the hopping parameters $t' \neq t$. d) Experimental evolution of the DOS with a uniaxial deformation of a honeycomb lattice with armchair boundaries. The merging transition occurs for a critical value $t' = 2t - 3t_3$. Under strain, new edge states appear at the band centre. e) These new edge states are located at the edges which are not parallel to the strain axis. (Figures adapted from [6].)

axis (Figure 6-e). (ii) Their localization along the edge increases when β increases. (iii) Their existence is not related to the topological transition: they appear as soon as $\beta > 1$. Moreover, it is found that (i) the intensity on one triangular sublattice stays zero, and (ii) the intensity on the other sublattice decreases roughly as $1/\beta^{2r}$, where r is the distance to the edge in units of the lattice parameter. These features are in agreement with the prediction of the existence of armchair edge states in deformed structures, and the existence of edge states has been related to a topological property of the bulk wave functions, the Zak phase [43]. More extensive investigation of these states, as well as of the states along zig-zag and bearded edges in anisotropic structures is presented in [44].

7. More Dirac points

In the framework of the general tight-binding model (5), one can also be confronted with situations where there are several pairs of Dirac points. The generation and motion of these additional Dirac points, as well as the possible fusion of Dirac points with like topological charge, are the issue of the present section.

7.1. Monolayer with third-neighbour coupling

In order to obtain additional pairs of Dirac points, the condition $f_{\vec{D}} = 0$ necessarily implies more harmonics in the dispersion relation [28, 45]. This can be achieved quite easily, at least in the framework of a toy model, by adding a third-nearest-neighbour coupling t_3 in the tight-binding model of graphene. We do not consider the coupling between second nearest neighbours which, by coupling sites of the same sublattice, modifies the dispersion relation but does not affect the existence of Dirac points, as long as the inversion symmetry is respected.[8] The Hamiltonian maintains the form (5), with the function $f_{\vec{k}}$ given by [11] (here $\beta = 1$),

$$f_{\vec{k}} = t(\beta + e^{i\vec{k}\cdot\vec{a}_1} + e^{i\vec{k}\cdot\vec{a}_2}) + t_3(e^{i\vec{k}\cdot(\vec{a}_1+\vec{a}_2)} + e^{i\vec{k}\cdot(\vec{a}_1-\vec{a}_2)} + e^{i\vec{k}\cdot(\vec{a}_2-\vec{a}_1)}). \qquad (53)$$

In graphene, the t_3 term is small. However, it is of interest to imagine a larger value of this parameter because it has a quite interesting effect on the evolution of the spectrum, as has been theoretically considered in Refs. [46, 47]. When t_3 increases and reaches the critical value $t/3$, a new pair of Dirac points emerges from each of the three inequivalent $\mathbf{M}$ points in reciprocal space (see Figure 7-a, b), following precisely the above universal scenario. As mentioned in Section 5, the annihilation as well as the emergence of Dirac points occurs necessarily at TRIM that are precisely the $\mathbf{M}$ points at the border of the hexagonal BZ between the K and K' points. Writing $\vec{k} = \vec{M} + \vec{q}$, we recover the Universal Hamiltonian in terms of the continuum wave vector $\vec{q}$ in the vicinity of $t_3 = t/3$ (keeping the leading-order terms) where the parameters m^*, c, Δ_* can be related to the original band parameters: $\Delta_* = t - 3t_3$, $c = 2ta$ and $m^* = 2/ta^2$. The parameter Δ_*,

---

[8]A coupling t_2 between second nearest neighbours dissymetrizes the spectrum. Interestingly, above a critical value $t_2 = t/6$, there is a $1/\sqrt{\text{energy}}$ Van Hove singularity at the band edge [42].

when it becomes negative $(t_3 > t/3)$, drives the emergence of a new pair of Dirac points at the **M**-point (Figure 7-b, c). The distance between the new Dirac points is given by $2q_D = 2\sqrt{-2m^*\Delta_*} = 4\sqrt{3t_3/t - 1}/a$.

We have thus added *three pairs* of Dirac points, each pair emerging from one of the three **M** points. When increasing further t_3, the new Dirac points approach the **K** and **K**$'$ points, so that each initial Dirac point sitting at the **K**$^{(')}$ points is now surrounded by three Dirac points (with opposite charges, see Figure 7.c.). These Dirac points merge at the critical value $t_3 = t/2$, and the spectrum becomes quadratic around **K**$^{(')}$ (Figure 7-b, c) [46, 47].

Near $t_3 \simeq t/2$, the Hamiltonian takes a new form (keeping leading-order terms) in the vicinity of the **K**$'$ point

$$\mathcal{H}'(\vec{q}) = \begin{pmatrix} 0 & -\dfrac{q^2}{2m^*} + c\,q^\dagger + \Delta \\ -\dfrac{q^{\dagger 2}}{2m^*} + c\,q + \Delta^* & 0 \end{pmatrix}, \tag{54}$$

where $q = q_x + iq_y$, and we have introduced by hand the gap parameter Δ for future discussions. The Hamiltonian in the vicinity of the **K** point is obtained by the substitution $q \rightarrow -q^\dagger$. Starting from Eqs. (5) and (53), we find $m^* = 4/9ta^2$, $c = 3(t_3 - t/2)a$, and $\Delta = 0$. This low-energy Hamiltonian is that of bilayer graphene [48], and one obtains moreover a parabolic band-contact point when $c = 0$. Within a tight-binding model, bilayer graphene is characterized essentially by three hopping integrals, the coupling γ_0 between nearest neighbours in each layer,[9] the coupling γ_1 between sites from different layers which are on top of each other, and the coupling γ_3 between nearest-neighbour sites from different layers which do not face each other. Neglecting this third coupling, the quadratic low energy spectrum in each valley is described by a 2×2 Hamiltonian of the form (54) with $c = \Delta = 0$, and a mass given by $m^* = 2\gamma_1/(9\gamma_0^2)$. For $\Delta = c = 0$, the eigenstates of Hamiltonian (54) are given by the same expression as those in equation (13), if one replaces $\phi_{\vec{q}} \rightarrow 2\phi_{\vec{q}}$. One thus notices that the associated winding number around a parabolic band-contact point is $w = \pm 2$.

The effect of the small γ_3 term is to induce a trigonal warping, so that the spectrum is no longer quadratic but consists of *four* Dirac points. This is in agreement with the additivity of topological charges discussed in Section 3.2 – indeed, the parabolic band-contact point with $w = 2$ is split into a central Dirac point with $w_{centr} = -1$ and three additional Dirac points with $w_i = +1$, such that the sum gives again $w = w_{centr} + 3w_i = 2$. This trigonal warping is described by Hamiltonian (54) with $c = -3\gamma_3/2$. The low-energy Hamiltonian for bilayers is thus equivalent to the Hamiltonian (53) of the single layer with third-nearest-neighbour coupling, the correspondance being $t \leftrightarrow 2\gamma_0^2/\gamma_1 + \gamma_3$ and $t_3 \leftrightarrow \gamma_0^2/\gamma_1$ [47].

---

[9]In order to make a clear distinction between the different parameters of equation (53) and the hopping integrals relevant for graphitic systems, we use here a notation in terms of γ_0 (named t in the previous sections), γ_1 and γ_3.

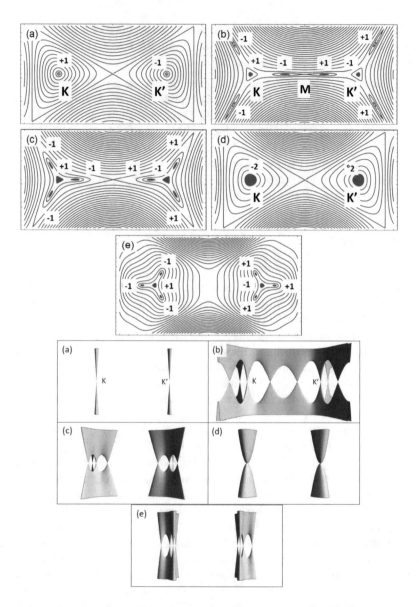

FIGURE 7. Top: Iso-energy lines in the vicinity of the $\mathbf{K}^{(')}$ and $\mathbf{M}$ points in the $t-t_3$ model [Eqs. (5) and (53)], for different values of the parameter t_3, (a) $t_3 = 0$, (b) $t_3 = 0.35t$, (c) $t_3 = 0.40t$, (d) $t_3 = 0.5t$, (e) $t_3 = 0.65t$. The vicinity of the Dirac points is indicated in red, as well as their associated winding number (defined in Section 3.2). Bottom: three-dimensional plot of the low energy spectrum for the same parameters.

7.2. Manipulation of Dirac points in twisted bilayer: a second type of merging

Twisted bilayer graphene consists of two graphene layers that have a rotational mismatch with respect to the conventional Bernal AB stacking. In order to understand its low-energy spectrum, consider for the moment two uncoupled layers that are rotated by a small angle θ with respect to the AB-stacking reference. In this case, the two Dirac cones associated with the two layers are separated in reciprocal space by a wave vector κ that is a function of θ. Numerical calculations indicate that no gap is opened at the Fermi level when interlayer hopping is taken into account [49]. However, the form of the interlayer coupling fixes the relative winding number of one Dirac cone with respect to the other one [50], and for small twist angles θ this coupling is continuously connected to that in the ideal AB case. The two Dirac points therefore have the same winding number, and, from a topological point of view, twisted bilayer at small angles is in the same class as AB-stacked bilayer graphene. Indeed, it can be described by the Hamiltonian

$$
\mathcal{H}_{++}(\vec{q}) = \frac{1}{2m^*} \begin{pmatrix} 0 & \dfrac{\kappa^2}{4} - q^2 \\ \dfrac{\kappa^{*2}}{4} - q^{\dagger 2} & 0 \end{pmatrix},
\tag{55}
$$

where the wave-vector shift is related to the gap parameter $\Delta = \kappa^2/8m^*$ of Hamiltonian (54), with $c = 0$ [50, 26]. A finite value of κ thus splits the quadratic dispersion relation into two cones separated by a saddle point (Figure 4-b). This Hamiltonian describes the merging of two Dirac points with the *same* charge and has to be contrasted with the Hamiltonian (46) which describes the merging of Dirac points with *opposite* charges. In contrast to the latter case, discussed in Section 5, there is no annihilation of the topological charges associated with the two Dirac points since one has a topological transition from $w_1 = +1$ and $w_2 = +1$ (for $\kappa \neq 0$) to $w = +2$ at the merging. The associated zero-energy Landau level in a magnetic field therefore remains two-fold degenerate from a topological point of view (in addition to the usual four-fold spin-valley degeneracy) regardless of the value of κ [50], a scenario that has recently been verified experimentally [51]. We notice finally that the most general situation with $c \neq 0$ and $\Delta \neq 0$ has been studied in Refs. [26, 52, 53, 54], in the framework of bilayer graphene, where one layer is displaced by a constant vector with respect to the other one, with no twist ($\theta = 0$).

8. Conclusions

In conclusion, we have discussed the basic properties of Dirac points that may occur in 2D crystalline systems, as well as their motion and merging. The physical systems that display such Dirac points involve, apart from mono- and bi-layer graphene, graphene-like systems, such as cold atoms in optical lattices, spatially modulated semi-conductor heterostructures, quasi-2D organic crystals, microwave lattices, molecular lattices, etc. Instead of an exhaustive discussion of all these artificial graphenes, we have illustrated the theoretical aspects of Dirac-point motion in only some of them. From the theoretical point of view, we have discussed some conditions for the emergence of Dirac fermions in generic two-band models as well as the role of discrete symmetries, such as time-reversal and inversion symmetry. Furthermore, we have aimed at a classification of the different types of Dirac-point merging, within a description of "second-generation" low-energy models and with the help of a topological analysis in terms of winding numbers. These winding numbers, which are revealed in the relative phase between the two components of the spinorial wave function, may be interpreted as topological charges. Very much as electric charges, the winding numbers are additive quantities, and their sum remains preserved in the different merging scenarios. Whereas the merging of Dirac points with opposite winding number, such as in the case of time-reversal symmetry related Dirac points, gives rise to a zero topological charge with the successive annihilation of the Dirac points and the opening of a gap in the spectrum, the situation is strikingly different in the case of Dirac-point merging with like topological charge. Indeed, in this case, the band-contact points are preserved because of a non-zero winding number. The set of parameters that give rise to a single (parabolic) band contact is singular in the sense that a slight change in the parameters splits the parabolic contact point into two Dirac points, as for example in the case of twisted bilayer graphene. The splitting into more than two Dirac points is also possible, albeit with a sum of ± 2 for the global topological charge, and occurs for instance in bilayer graphene at very low energies, where trigonal warping becomes visible – in this case, one finds a central Dirac point with a winding number $w = -1$ surrounded by three satellite Dirac points with $w = +1$ (for a total charge of 2, here).

Apart from the stability of the band-contact points, the topological charges also allow us to understand other physical quantities. They are revealed under the influence of a magnetic field applied perpendicular to the 2D system. This field quantizes the particles' energy into discrete Landau levels that are highly degenerate from an orbital point of view. In addition to this orbital degeneracy and the spin degree of freedom, one finds a topological degeneracy of the zero-energy level that is precisely related to the topological charge. Whereas Dirac points that are far apart in reciprocal space (as compared to the inverse magnetic length) provide each a zero-energy Landau level, as stipulated by the low-energy Dirac-fermion model, the situation becomes more complicated in the vicinity of the merging transitions. On the one hand, the merging of time-reversal symmetry related Dirac points (with a zero total winding number) destroys the (originally two-fold valley-degenerate) zero-energy level and splits it into two seperate levels. On the other hand, the merging of two Dirac points with the same winding number has no effect on the zero-energy level, which thus remains two-fold degenerate. The topological charge (the total winding number) therefore indicates the number of topologically protected zero-energy levels, as compared to the total number of possible zero-energy levels, which coincides with the total number of Dirac points in the absence of a magnetic field.

Acknowledgment

The work presented here results from several fruitful collaborations. First of all, we would like to acknowledge the long-term in-house collaboration with our colleagues J.-N. Fuchs and F. Piéchon. Furthermore we would like to thank our students and postdocs R. de Gail, P. Delplace, P. Dietl, and L.-K. Lim, as well as our external collaborators M. Bellec, U. Kuhl, F. Mortessagne, F. Guinea, and A.H. Castro Neto.

References

[1] For a review, see Castro Neto, A.H., Peres, N.M.R., Novoselov, K.S., and Geim, A.K.: Rev. Mod. Phys. **81**, 109 (2009).

[2] Hasan, M.Z., and Kane, C.L.: Rev. Mod. Phys. **82**, 3045 (2010); Qi, X.-L., and Zhang, S.-C.: Rev. Mod. Phys. **83**, 1057 (2011); Carpentier, D.: Topology of Bands in Solids: From Insulators to Dirac Matter. pp. 87–118, in this volume.

[3] Katayama, S., Kobayashi, A., and Suzumura, Y.: J. Phys. Soc. Jap. **75**, 054705 (2006); Kobayashi, A., Katayama, S., Suzumura, Y., and Fukuyama, H.: J. Phys. Soc. Jap. **76**, 034711 (2007); Goerbig, M.O., Fuchs, J.N., Piéchon, F., and Montambaux, G.; Phys. Rev. B **78**, 045415 (2008).

[4] Tarruell, L., Greif, D., Uehlinger, T., Jotzu, G., and Esslinger, T.: Nature **483**, 302 (2012).

[5] Gomes, K.K., Mar, W., Ko, W., Guinea, F., and Manoharan, H.C.: Nature **483**, 306 (2012).

[6] Bellec, M., Kuhl, U., Montambaux, G., and Mortessagne, F.: Phys. Rev. Lett. **110**, 033902 (2013).

[7] Hosur, P., and Qi, X.-L.: C. R. Physique **14**, 857 (2013).

[8] Nielsen, H.B.: Nucl. Phys. B **185**, 20 (1981).

[9] Berry, M.V.: Proc. R. Soc. Lond. A **392**, 45 (1984).

[10] Wallace, P.R.: Phys. Rev. **71**, 622 (1947).

[11] About the writing of the function $f_{\vec{k}}$, see Bena, C., and Montambaux, G.: New J. Phys. **11**, 095003 (2009).

[12] Abergel, D.S.L., Apalkov, V., Berashevich, J., Ziegler, K., and Chakraborty, T.: Adv. Phys. **59**, 261 (2010).

[13] Peres, N.M.R.: Rev. Mod. Phys. **82**, 2673 (2010).

[14] Das Sarma, S., Adam, S., Hwang, E.H., and Rossi, E.: Rev. Mod. Phys. **83**, 407 (2011).

[15] Novoselov, K.S.: Rev. Mod. Phys. **83**, 837 (2011); Geim, A.K.: Rev. Mod. Phys. **83**, 851 (2011).

[16] de Heer, W.A., Berger, C., Wu, X., Sprinkle, M., Hu, Y., Ruan, M., Stroscio, J.A., First, P.N., Haddon, R., Piot, B., Faugeras, C., Potemski, M., and Moon, J.-S.: J. Phys. D: Appl. Phys. **43**, 374007 (2010).

[17] Kotov, V.N., Uchoa, B., Peirera, V.M., Castro Neto, A.H., and Guinea, F.: Rev. Mod. Phys. **84**, 1067 (2012).

[18] Goerbig, M.O.: Rev. Mod. Phys. **83**, 1193 (2011).

[19] McClure, J.W.: Phys. Rev. **104**, 666 (1956).

[20] Sadowski, M.L., Martinez, G., Potemski, M., Berger, C., and de Heer, W.A.: Phys. Rev. Lett. **97**, 266405 (2006); Jiang, Z., Henriksen, E.A., Tung, L.C., Wang, Y.-J., Schwartz, M.E., Han, M.Y., Kim, P., and Stormer, H.L.: Phys. Rev. Lett. **98**, 197403 (2007).

[21] Li, G., and Andrei, E.: Nature Phys. **3**, 623 (2007).

[22] Novoselov, K.S., Geim, A.K., Morosov, S.V., Jiang, D., Katsnelson, M.I., Grigorieva, I.V., Dubonos, S.V., and Firsov, A.A.: Nature **438**, 197 (2005).

[23] Zhang, Y., Tan, Y.-W., Stormer, H.L., and Kim, P.: Nature **438**, 201 (2005).

[24] Goerbig, M.O.: *Quantum Hall Effects*. Lecture notes of the Les Houches Summer School 2009 (Singapore Session), C. Miniatura, L.-C. Kwek, M. Ducloy, B. Grémaud, B.-G. Englert, L. Cugliandolo, A. Ekert, and K.K. Phua (Eds.), Oxford UP (Oxford 2011).

[25] Onsager, L.: Phil. Mag. **43**, 1006 (1952); Lifshitz, I.M., and Kosevich, A.M.: Sov. Phys. J.E.T.P. **2**, 636 (1956).

[26] de Gail, R., Goerbig, M.O. and Montambaux, G.: Phys. Rev. B **86**, 045407 (2012).

[27] Koghee, S., Lim, L.-K., Goerbig, M.-O., and Morais Smith, C.: Phys. Rev. A **85**, 023637 (2012).

[28] Montambaux, G., Piéchon, F., Fuchs, J.-N., and Goerbig, M.O.: Phys. Rev. B **80**, 153412 (2009); Eur. Phys. J. B **72**, 509 (2009).

[29] Dietl, P., Piéchon, F., and Montambaux, G.: Phys. Rev. Lett. **100**, 236405 (2008).

[30] Banerjee, S., Singh, R.R.P., Pardo, V., and Pickett, W.E.: Phys. Rev. Lett. **103**, 016402 (2009).

[31] Hasegawa, Y., Konno, R., Nakano, H., and Kohmoto, M.: Phys. Rev. B **74**, 033413 (2006).

[32] Wunsch, B., Guinea, F., and Sols, F.: New J. Phys. **10**, 103027 (2008).

[33] Pereira, V.M., Castro Neto, A.H., and Peres, N.M.R.: Phys. Rev. B **80**, 045401 (2009).

[34] Rechtsman, M.C., Plotnik, Y., Zeuner, J.M., Szameit, A., and Segev, M.: Phys. Rev. Lett. **111**, 103901 (2013).

[35] Kuhl, U., Barkhofen, S., Tudorovskiy, T., Stockmann, H.-J., Hossain, T., de Forges de Parny, L., and Mortessagne, F.: Phys. Rev. B **82**, 094308 (2010).

[36] Bittner, S., Dietz, B., Miski-Oglu, M., Oria Iriarte, P., Richter, A., and Schaefer, F.: Phys. Rev. B **82**, 014301 (2010).

[37] Jacqmin, T., Carusotto, I., Sagnes, I., Abbarchi, M., Solnyshkov, D., Malpuech, G., Galopin, E., Lemaître, A., Bloch, J., and Amo, A.: Phys. Rev. Lett. **112**, 116402 (2014).

[38] Polini, M., Guinea, F., Lewenstein, M., Manoharan, H.C., and Pellegrini, V.: Nature Nanotech. **8**, 625 (2013).

[39] Lim, L.-K., Fuchs, J.-N., and Montambaux, G.: Phys. Rev. Lett. **108**, 175303 (2012); Fuchs, J.-N., Lim, L.-K., and Montambaux, G.: Phys. Rev. A **86**, 063613 (2012).

[40] Landau, L.D.: Phys. Z. Sow. **2**, 46 (1932); Zener, C.: Proc. R. Soc. London A **137**, 696 (1932); see also Wittig, C.: J. Phys. Chem. B **109**, 8428 (2005).

[41] Lim, L.-K., Fuchs, J.-N., and Montambaux, G.: Phys. Rev. Lett. **112**, 155302 (2014).

[42] Bellec, M., Kuhl, U., Montambaux, G., and Mortessagne, F.: Phys. Rev. B **88**, 115437 (2013).

[43] Delplace, P., Ullmo, D., and Montambaux, G.: Phys. Rev. B **84**, 195452 (2011).

[44] Bellec, M., Kuhl, U., Montambaux, G., and Mortessagne, F.: New J. Phys. **16**, 113023 (2014).

[45] Sticlet, D., and Piéchon, F.: Phys. Rev. B **87**, 115402 (2013).

[46] Bena, C., and Simon, L.: Phys. Rev. B **83**, 115404 (2011).

[47] Montambaux, G.: Eur. Phys. J. B **85**, 375 (2012).

[48] McCann, E., Falko, V.I.: Phys. Rev. Lett. **96**, 086805 (2006).

[49] Trambly de Laissardière, G., Mayou, D., and Magaud, L.: Phys. Rev. B **86**, 125413 (2012).

[50] de Gail, R., Goerbig, M.O., Guinea, F., Montambaux, G., and Castro Neto, A.H.: Phys. Rev. B **84**, 045436 (2011).

[51] Lee, D.S., Riedl, Ch., Beringer, T., Castro Neto, A.H., von Klitzing, K., Starke, U., and Smet, J.H.: Phys. Rev. Lett. **107**, 216602 (2011); Sanchez-Yamagishi, J.D., Taychatanapat, T., Watanabe, K., Taniguchi, T., Yacoby, A., and Jarillo-Herrero, P.: Phys. Rev. Lett. **108**, 076601 (2012).

[52] Son, Y.-W., Choi, S.-M., Hong, Y.-P., Woo, S., and Jhi, S.-H.: Phys. Rev. B **84**, 155410 (2011).

[53] Mucha-Kruczynski, M., Aleiner, I.L., and Fal'ko, V.I.: Solid State Comm. **151**, 1088 (2011).

[54] de Gail, R., Fuchs, J.-N., Goerbig, M.O., Piéchon, F., and Montambaux, G.: Physica B **407** 1948 (2012).

Mark Goerbig and Gilles Montambaux
Laboratoire de Physique des Solides – Bât. 510
Université Paris Sud
CNRS UMR 8502
F-91405 Orsay cedex, France
e-mail: mark-oliver.goerbig@u-psud.fr
          gilles.montambaux@u-psud.fr

Dirac Matter, 55–73
© 2016 Springer Basel AG

Quantum Transport in Graphene: Impurity Scattering as a Probe of the Dirac Spectrum

Chuan Li, Sophie Guéron and Hélène Bouchiat

Abstract. Since the very first investigations of the electronic properties of graphene, the nature of the scattering disorder potential has been shown to play an essential role in determining the carrier density dependence of the conductance. Impurity scattering is characterized by two different times, the transport and elastic scattering times, which are sensitive to the particular Dirac spectrum of graphene. The analysis of the ratio between these two times gives insight on the nature (neutral or charged) and range of the scatterers. We show how to extract these two times from magneto-transport measurements and analyze their differences in monolayer and bilayer Graphene in relation with the different symmetry properties of their band structure and wave functions. It is found that whereas short range impurity scattering is the dominant mechanism limiting the classical transport, phase coherent mesoscopic transport is very sensitive to long range disorder. In particular, the formation of electron/hole puddles in the vicinity of the charge neutrality point strongly affects the transport of Andreev pairs in the presence of superconducting electrodes. We will also discuss the modification of electronic properties of graphene in the presence of adsorbed atoms and molecules and in particular focus on spin dependent scattering on adsorbates leading to a spin orbit interaction. There is indeed a big interest in controlling and inducing spin orbit interactions in graphene. One can hope to induce and detect a spin Hall effect with a great potential impact in graphene based spintronic devices and ultimately reach a regime of quantum spin Hall physics. In contrast with these very short range scatterers, we discuss the possibility to engineer networks of longer range strained regions in which electronic properties are locally modified by transferring graphene on arrays of silicon oxyde nanopillars.

1. Introduction

Graphene, the single layer of graphite, is one of the most simple and intriguing two-dimensional conducting material [1]. The low energy electronic spectrum of graphene is constituted by two Dirac cones centered around the two non equivalent K points of the Brillouin zone with a perfect electron-hole symmetry at the Fermi level [2]. The electronic wave functions are characterized by their pseudo spin components on the two nonequivalent A and B atoms of the hexagonal lattice. The symmetry properties of this lattice give rise to a relative phase factor between the 2 spinorial components which depends on the orientation of the Bloch wave vector. This phase factor determines the sensitivity of electronic transport to the disorder potential produced by impurities or charge inhomogeneities. In particular it was emphasized that backward scattering within one valley is forbidden by these symmetry properties of the wave functions which strongly suppresses Anderson localization in Graphene. The same effect exists also in carbon nanotubes [3]. This is however not true in the presence of very short range impurities such as single atom vacancies which induce intervalley scattering.

One way to tackle the scattering mechanism in graphene is to compare the two different scattering times: (i) the transport time τ_{tr}, which governs the current relaxation and enters the Drude conductivity (σ), (ii) the elastic scattering time τ_e, which is the lifetime of a plane wave state [4]. Since τ_{tr} and τ_e involve different angular integrals of the differential cross section, they differ as soon as the Fourier components of the potential depend on q. A large ratio τ_{tr}/τ_e indicates that scattering is predominantly in the forward direction, so that transport is not affected much by this type of scattering. This is the case in 2D electron gases (2DEG) confined in GaAs/GaAlAs heterojunctions with the scattering potential produced by remote charged Si donors [5], where τ_{tr}/τ_e is larger than 10. We show how to extract these two times from magnetotransport. These experiments are discussed in the first part of the paper. The ratio τ_{tr}/τ_e is found to be close to 1.8 and nearly independent of the carrier density. Comparison with theoretical predictions suggests that the main scattering mechanism which dominates the physics of classical transport between 1 and 10 K is due to strong (resonant) scatterers of a range shorter than the Fermi wavelength.

The second part of the paper is devoted to quantum mesoscopic transport at very low temperature in phase coherent samples. It is dominated by interference effects between all wave packets crossing the sample. This interference pattern is sensitive to variations in disorder configuration, Fermi energy or magnetic flux, leading to reproducible conductance fluctuations as one of these parameters is changed.

These fluctuations which are the trade mark of mesoscopic transport have been investigated in mesoscopic graphene samples and characterized them by their correlations and amplitudes as a function of Fermi energy and magnetic field. This analysis has revealed a strong sensitivity of conductance fluctuations to the long range disorder induced by the formation of charge inhomogeneities (electron hole puddles) [6]. Another way to probe the coherent nature of quantum transport in mesoscopic systems is to investigate proximity induced superconductivity in the presence of superconducting electrodes. In particular the induced Josephson supercurrent has been shown to decay on the thermal and phase coherence lengths scales L_T and L_φ characteristic of mesoscopic transport. Supercurrents could be successfully measured in short graphene Josephson corresponding to $L < \xi_S$ where ξ_S is the superconducting coherence length [7, 8, 9, 10]. We present also measurements in the long junction limit $L >> \xi_S$. Whereas a supercurrent is clearly observed far from the neutrality point, its value is strongly suppressed close to the neutrality point. These results suggest the low carrier-density puddles, along which specular Andreev reflection occurs, play a crucial role in the supercurrent suppression.

The third part of this paper is essentially prospective. Beyond the investigation of impurity scattering in graphene, we discuss the possibility of designing new functionalities in graphene by controlling the nature of intentionally deposited scattering centers. We will first consider the case of adatoms, small aggregates or molecules chosen for their high spin orbit scattering properties. The main motivation is to induce spin orbit interactions in graphene. One can hope to induce and detect a gate dependent spin Hall effect with a great potential impact in graphene based spintronic devices and ultimately reach a regime of quantum spin Hall physics. It has recently been shown that the measurements of transport and elastic times discussed in Section 2 provide a quantitative determination of the spin orbit strength of added scatterers. Following the experiments of Crommie et al. [11] which demonstrated that local strain induces a polarization of the pseudo spin degrees of freedom acting as an effective magnetic field, we will show that it is also possible to create local strained regions by deposition of graphene on etched substrates consisting of arrays of sharp silicon oxide nanopillars. We expect electronic properties to be strongly modified in the strained graphene regions with an increase of the density of states at the Dirac point.

2. Impurity scattering in graphene: determination of the transport and elastic scattering times

The nature of impurity scattering in graphene has been the subject of intense debate these last years. In particular the quasi-linear carrier density dependence of the conductivity $\sigma(n_c)$ cannot be understood when simply considering neutral short range impurities with "white noise" (wavevector q independent) scattering: indeed the associated inverse scattering time varies linearly with the density of states leading to a conductivity independent of the carrier density. Including inter valley scattering [3, 12, 13] leads only to a weak (logarithmic) dependence of $\sigma(n_c)$. In contrast, scattering on charged impurities originates from a q dependent screened Coulomb potential described in the Thomas–Fermi approximation [14, 15, 16]. This leads to a linear $\sigma(n_c)$. However experiments performed to probe this question, that measured the change in σ upon immersion of graphene samples in high dielectric constant media, found a dependence of the conductance with the dielectric constant which is much too weak to be understood by charged impurity scattering [17], see also the more recent work on SrTiO$_3$ substrates [18]. Alternate explanations involve resonant scattering centers with a large energy mismatch with the Fermi energy of carriers [13, 19]. We will see that these strong scattering centers play indeed the major role in the determination of the classical conductivity. Impurity scattering is characterised by the transport and elastic scattering times which are related to the following angular integrals of the scattering crossection $\sigma(\theta)$:

$$1/\tau_e = n_{imp}v_F \int d\theta\sigma(\theta),$$

$$1/\tau_{tr} = n_{imp}v_F \int d\theta(1 - \cos\theta)\sigma(\theta). \tag{1}$$

For isotropic relatively short range scatterers of range R smaller than the Fermi wavelength but longer than the interatomic distance a which forbids intervalley scattering, the angular dependence of the scattering cross section $\sigma(\theta)$ contains an angular dependence in $\sigma_0\frac{1+\cos j\theta}{2}$. The Berry phase factor $j\theta$ is specific to the symmetry of the wave functions within a single valley. One has $j = 1$ for the graphene monolayer leading to the absence of backscattering which is also at the origin of Klein tunneling. On the other hand $j = 2$ for the bilayer. As a result from equation (1) one expects the ratio τ_{tr}/τ_e to be equal to 2 for the monolayer and 1 for the bilayer.

In the following we discuss how these transport and elastic scattering times can be determined from magnetoresistance experiments. In high magnetic field, when the cyclotron frequency is larger than $1/\tau_e$, the magneto-conductivity exhibits Shubnikov–de Haas (ShdH) oscillations related to the formation of Landau levels. The broadening of these levels at low temperature yields τ_e, while the low field quadratic magneto-conductivity yields τ_{tr}. The samples investigated were fabricated by exfoliation of natural graphite flakes and deposition on a doped silicon substrate with a 285 nm thick oxide. The carrier density can be tuned from electrons to holes through the charge neutrality point by applying a voltage on the backgate. The ML and BL samples were identified using Raman spectroscopy. The electrodes were fabricated by electron beam lithography and either sputter deposition of 40 nm thick palladium or Joule evaporation of a bilayer 5nmTi/70nm Au. The low temperature magnetoresistance (MR) is shown on Figure 1 for 2 different samples, a monolayer and a bilayer. It exhibits a quadratic behavior at low field whereas Shubnikov–de Haas oscillations appear at high field. We now describe how we extract τ_{tr} and τ_e from this data (see Figure 1). The two-terminal MR results from mixing of the diagonal (ρ_{xx}) and off-diagonal (ρ_{xy}) components of the resistivity tensor [21, 22]. The degree of mixing depends on the aspect ratio of the sample. For a square geometry, close to that of the monolayer, $R(B) = \left[\rho_{xx}^2 + \rho_{xy}^2\right]^{1/2}$; in a short wide sample such as the bilayer $R(B) = (L/W)\left[\rho_{xx}^2 + \rho_{xy}^2\right]/\rho_{xx}$. Intermediate geometries can be calculated following the model developed in [21]. It is then possible to reconstruct the complete MR from the expressions of the resistivity tensor [23] valid in the limit of moderate magnetic field where ShdH oscillations can be approximated by their first harmonics:

$$\delta\rho_{xx}(B)/\rho_0 = 4D_T\exp\left[-\frac{\pi}{\omega_c\tau_e}\right]\cos\left[\frac{j\pi E_F}{\hbar\omega_c} - \phi\right]$$

$$\rho_{xy}(B) = \rho_0\omega_c\tau_{tr} - \delta\rho_{xx}(B)/2\omega_c\tau_{tr}, \tag{2}$$

where $\rho_0 = 1/\sigma$ is the zero-field resistivity and $\omega_c = eB/m^*$ is the cyclotron frequency, $m^* = \hbar k_F/v_F$ is the cyclotron mass which depends explicitly on the Fermi wave vector k_F for the ML (constant Fermi velocity). On the other hand, the bilayer's dispersion relation is parabolic at low energy and m^* can be approximated by the effective mass $m_{\mathrm{eff}} = 0.035 m_e$, nearly independent of the carrier density in the range of V_g explored where $|E_F| \leq 80$ meV is smaller by a factor 5 than the energy band splitting [1]. The phase ϕ, either π or 2π, and the parameter j, either 1 or 2, depend on the nature of the sample (ML or BL). The Fermi energy E_F is $\hbar k_F v_F$ for the monolayer and $\hbar^2 k_F^2/(2m_{\mathrm{eff}})$ for the bilayer. The prefactor $D_T = \gamma/\sinh(\gamma)$ with $\gamma = 2\pi^2 k_B T/\hbar\omega_c$ describes the thermal damping of the oscillations.

To analyze the data we first deduce k_F from the periodicity of the ShdH oscillations function of $1/B$. This determination is more reliable than the estimation of $n_c = k_F^2/\pi$ from the gate voltage and the capacitance between the doped silicon substrate and the graphene sample. This is specially the case close to the neutrality point where the sample is possibly inhomogeneous [20]. Knowing k_F we then determine τ_{tr} from the low field quadratic magnetoresistance which is found to be independent of temperature between 1 and 4 K:

$$R(B) - R(0) = \frac{h}{2e^2}\frac{L}{W}\frac{1}{k_F v_F \tau_{tr}}\alpha_g(\omega_c \tau_{tr})^2. \tag{3}$$

We have used the relation $\sigma = \rho_0^{-1} = (2e^2/h)k_F v_F \tau_{tr}$. The dimensionless coefficient α_g, which depends on the aspect ratio of the sample was determined numerically following [21] and the experimental values of W and L. It is important to note that this determination of τ_{tr} is independent of any assumption of the contact resistance on two terminal samples. We finally extract τ_e from the damping of the first harmonic of ShdH oscillations in the resistivity tensor in $\exp(-\beta/B)$ where $\beta = \pi\hbar k_F/ev_F \tau_e$, see equation (2).

The k_F dependences of τ_{tr} and τ_{tr}/τ_e are shown in Figure 3 for several two terminal samples A, B and C and two multi-terminal samples (D and E) with Hall-bar geometry (see [24] for more details). We observe different behaviors for the ML samples, where τ_{tr} has a minimum at the CNP, and the BL, where it has a maximum. In all cases, despite rather large variations of τ_{tr}, τ_{tr}/τ_e is nearly independent of k_F. It is equal to 1.7 ± 0.3 for the monolayers A,C,E and to 1.8 ± 0.2 for the bilayer in the whole range explored, which corresponds to n_c between 1.5×10^{11} and 5×10^{12}cm^{-2}. As already discussed, τ_{tr}/τ_e of the order but smaller than 2 indicates that the typical size of the scatterers does not exceed the Fermi wavelength for the ML samples. According to equation (1), the factor 2 instead of 1 expected for short range impurities in GaAs is due to the symmetry of the wave functions within one valley, more precisely to the π Berry phase factor which forbids backscattering within one valley. A smaller ratio would have been however expected for the BL, since symmetry of wave functions do not forbid intravalley backscattering in that case. We will see in the following that this ratio

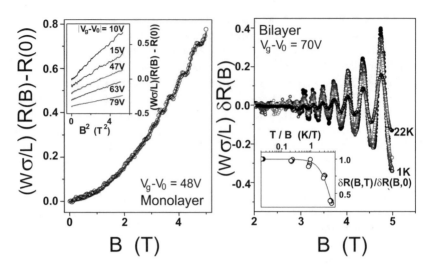

FIGURE 1. Analysis of the magnetoresistance. Left panel: Magnetoresistance of a monolayer (sample A). Dots: experimental points at $T = 1K$;
Continuous line: fit according to Eqs. (3) and (2). Inset: B^2 dependence
of the low-field magnetoresistance for different gate voltages (Curves
shifted along the y-axis for clarity). τ_{tr} is extracted from the slopes of
these curves according to equation (3). Notice that the slope increases
in the vicinity of the Dirac point reflecting the divergence of the inverse
effective mass. Right panel: ShdH oscillations of the longitudinal component of the resistivity in a bilayer sample B for different temperatures
after subtraction of the quadratic background. The Fermi wavevector
k_F and the elastic time τ_e are deduced from the period and the decay
of the oscillations with $1/B$ at low temperature. Inset: Temperature dependence of the oscillations amplitude normalized to $T = 0$. Solid line:
fit according to the Lifshitz–Kosevich formula $D_T = \gamma/\sinh(\gamma)$ with
$\gamma = 2\pi^2 k_B T/\hbar\omega_c$ [23]. The effective mass determined from this fit is
$m_{\mathrm{eff}} = 0.035 \pm 0.002 m_e$ in the whole range of gate voltage investigated.

τ_{tr}/τ_e can be used to probe the presence of spin orbit scattering impurities which
modify the symmetry of the wave functions in graphene.

We now compare our results on τ_e and τ_{tr} to theoretical predictions. We
first consider scattering on charged impurities [14, 15]. Screened charged impurities are characterized by a screening radius $1/q_{sc}$, which in the Thomas–Fermi
approximation, is given by $1/q_{TF} \equiv \pi\epsilon\hbar v_F/e^2 k_F$, where ϵ is the appropriate dielectric constant. In the Born approximation, the transport time is $\tau_{tr} \propto q_{sc}^2 v_F/k_F$.
For a monolayer, q_{TF}/k_F is a constant $\simeq 3$ and both τ_{tr} and τ_e are then expected to increase as k_F, which is not what we observe in Figure 3 where the
increase is sublinear. The disagreement is even stronger for a bilayer, where the

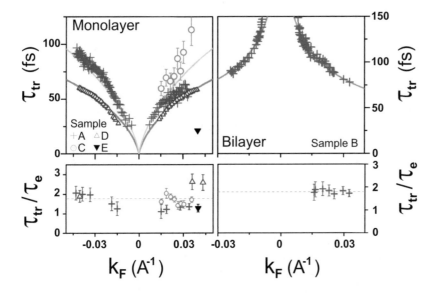

FIGURE 2. k_F dependence of τ_{tr} and τ_{tr}/τ_e ratio. Left panel: monolayers A, C, D and E. Right panel: bilayer B. The continuous lines are the fits for samples A, B and D according to the resonant impurity model, equation (4). For samples A B and C (two terminal configuration) τ_{tr} was extracted from the low field magnetoresistance (crosses) whereas it was extracted from the zero field conductivity for samples D and E. Positive/negative values of k_F correspond to electron/hole doping. Lower panels: ratio τ_{tr}/τ_e where τ_e is deduced from the fit of the low temperature decay of the ShdH oscillations. Dotted lines figure the average value $\tau_{tr}/\tau_e = 1.8$. We note that sample D exhibits a value of τ_{tr}/τ_e at high electron doping which is larger than 2 ($\simeq 2.4$). The area of this sample ($12\mu m^2$) is much larger than the area ($\simeq 1\mu m^2$) of all the other samples A, B, C and E. We suspect that this large sample contains more spatial inhomogeneities than the other smaller samples which could explain a reduced value of τ_e.

ratio $q_{TF}/k_F \propto 1/k_F$ varies between 3 at high doping and 12 close to the neutrality point. The transport time is then expected to vary linearly with n_c, if the screening radius is estimated as $\sim 1/k_F$, or to be independent of k_F if estimated as $\sim 1/q_{TF} \ll 1/k_F$ [15], neither of which agrees with our data, see Figure 3.

An alternative explanation is resonant scattering resulting from vacancies or any other kind of impurities of range R such that $a \lesssim R \ll 1/k_F$, where a is the carbon-carbon distance, and with a large potential energy [13, 19, 24]. It is characterized by a transport cross section

$$A_{tr} \simeq \pi^2 / \left(k_F \ln^2(k_F R) \right). \tag{4}$$

The resulting transport time $\tau_{tr} = 1/(n_i v_F A_{tr})$ (n_i is the concentration of impurities) leads to a conductance increasing as n_c with logarithmic corrections for both the ML and BL. In both cases, our extracted $\tau_{tr}(k_F)$ (see Figure 3) are compatible with the square logarithmic dependence of equation (4).

This analysis also corroborates our results on the ratio τ_{tr}/τ_e indicating scatterers with a range smaller than the Fermi wavelength (but possibly of the order of or slightly larger than the lattice spacing). Whereas the resonant character is not essential for the validity of equation (4) for massive carriers (corresponding to the bilayer) [26], it has been shown that it is essential for massless carriers in the monolayer [27]. This resonant-like character, although not straightforward has been demonstrated in the case of scattering centers created by vacancies in graphene over a wide range of Fermi energies [28]. As shown in detail in [24], it is not necessary to fine-tune k_F to obtain the $\ln^2$ dependence in equation (4).

In conclusion to this section, our results indicate that the main scattering mechanism which determines the classical transport properties in our graphene samples is due to strong neutral defects, with a range shorter than the Fermi wavelength and possibly of the order of a, inducing resonant (but not unitary) scattering. Likely candidates are vacancies, as observed in transmission electron microscopy [29], voids of several atomic size, ad-atoms or short range ripples as suggested in [30] [31]. It was pointed out that these ripples also give rise to intravalley dephasing with a characteristic time which can be deduced from weak localisation measurements [32].

3. Quantum transport: proximity induced superconductivity and specular Andreev reflection.

We now move to the low temperature transport measured below 1K in the phase coherent regime where the length of the samples investigated does not exceed the characteristic length for mesoscopic transport L_φ and L_T, the phase coherence and thermal length. We have chosen to focus on the situation where graphene is connected to superconducting electrodes giving rise to the formation of Andreev pairs which are coherent superposition of time reversed electron hole pairs. It was shown that the formation of these Andreev pairs could be strongly affected by the nature of the symmetric electron-hole band structure of graphene and be very sensitive to charge inhomogeneities in vicinity of the neutrality point [34]. Indeed, transport across a Superconductor/Normal metal (S/N) interface at subgap energy implies extracting two electrons from the superconductor and injecting them into the N, which produces a correlated Andreev pair in the normal metal. In a usual normal metal, which is highly doped in the sense that the Fermi level lies well within the conduction band, both electrons are injected in the conduction band of the N. The two injected members of the Andreev pair then follow the same, albeit time-reversed, diffusive path in the normal conductor, so that coherent propagation can occur over several micrometers (the phase coherence length at low temperature).

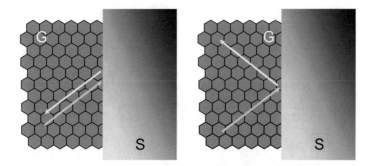

FIGURE 3. Sketch of the retro- and specular Andreev reflection at a G/S interface. Left: Retro-reflection occurs in usual conductors and in doped graphene, where the Fermi energy much exceeds the superconducting electrode's energy gap Δ, $E_F \gg \Delta$. Right: The specular Andreev reflection occurs in graphene at doping small enough that $E_F \ll \Delta$.

This coherent propagation leads to supercurrents that flow through such normal conductors several microns long connected to two superconductors. In contrast, at a superconductor/graphene (S/G) interface, if the superconductor's Fermi level is aligned with the graphene Dirac point, the two electrons of a Cooper pair must split into an electron in the conduction band and the other in the valence band. The two members of the injected pair in the graphene now have the same velocity (rather than opposite) parallel to the S/G interface (see Figure 3) and thus do not follow the same diffusive path. The observation of this special type of pair injection, also called "specular Andreev reflection", has so far remained elusive. This is because the doping inhomogeneities in the graphene samples, of several milli-electronvolts [33], are much larger than the superconductor's energy gap. Thus only the usual injection of counter-propagating electron pairs (also called Andreev retroreflection) sets in.

In the following we show that diffusive transport of Andreev pairs through quantum coherent graphene reveals an analog of specular Andreev reflection at an S/G interface, in the form of specular reflections of Andreev pairs at the interface between a doped charge puddle and a zero density region. These processes result in the destruction of counter-propagation upon specular reflection, and lead to a large phase accumulation within each Andreev pair. Since all pairs contribute to the supercurrent with their phase, the resulting supercurrent is suppressed. We argue that this specular reflection explains the suppression of the critical current that we observe near the Charge Neutrality Point (CNP) in quantum coherent, long and diffusive SGS junctions see Figure 4. Gate tunable Josephson junctions have been investigated by a large number of groups, we focus in the following on the behavior of long ($L \gg \xi_S$) compared to short ($L \ll \xi_S$) junctions where ξ_S is the superconducting coherence length. We compare data obtained on a long junction

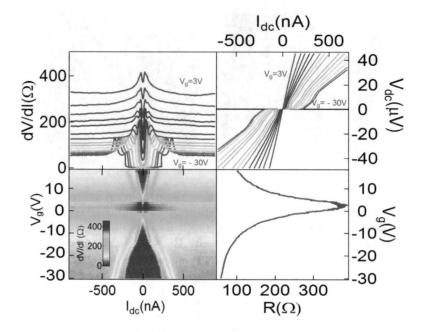

FIGURE 4. Proximity effect in graphene connected to Nb electrodes at 200 mK. Upper left panel: dV/dI vs I_{dc} for different gate voltages, and, bottom left panel, its two-dimensional color plot. The suppression of critical current in a gate voltage region of $\pm 10V$ around the charge neutrality point is noticeable. Upper right panel: I(V) curves for different gate voltages, showing how the proximity effect varies between a full proximity effect with zero resistance at high doping, and quasi normal behavior with a linear IV around the charge neutrality point. Lower right panel: Zero bias differential resistance as a function of gate voltage in the normal state, from which the R_N is determined. A small magnetic field was applied to destroy the constructive interference leading to the supercurrent. From Komatsu et al. [36].

$1\mu_m long$ graphene samples connected to Pd/Nb electrodes and a short junction 0.3 μm long with Ti/Al electrodes. In the theory of the proximity effect the critical current of a short junctions is $I_c = \Delta/eR_N$ whereas in the diffusive, long junction limit, the critical current has a maximum zero temperature value given by the Thouless energy E_{Th} divided by the normal resistance state R_N, multiplied by a numerical factor α which depends on the junction length L: $I_c = \alpha E_{Th}/eR_N$, where $E_{Th} = \hbar D/L^2$, with D is the diffusion constant.

The tunability of graphene is an asset to probe these relations. As shown in Figure 5, one can compare the measured switching current for the short and long junctions investigated to the superconducting gap (respectively the Thouless

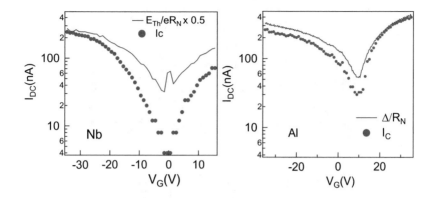

FIGURE 5. Comparison between the gate voltage dependence of the critical current of the long Nb/G/Nb junction with the gate voltage dependence of the critical current of the short Al/G/Al junction compared to the theoretical predictions. For the long junction we observe a strong suppression of the critical current close to the Dirac point. The critical current of the short junction follows the expected value Δ/R_N.

energy) divided by the normal state resistance as the gate voltage is varied. It is clear from Figure 5 that there is not a constant factor between E_{Th}/eR_N and I_c for the long junction but that I_c is strongly suppressed at small gate voltage, as the charge neutrality point is approached. Note that this suppression is not observed in the short SGS junction. It has not been reported in the other graphene based SNS junctions, measured in other groups, which are more than two or three times shorter than the long devices we have investigated.

We attribute this suppression close to the CNP to specular reflection of an Andreev pair at the charge puddle contours, as sketched in Figure 6. Indeed, around the CNP, electron-doped regions coexist with hole doped ones, forming a network of so-called puddles [33]. Where the doping varies from n to p doping there is necessarily a boundary with exactly zero doping, to within $k_B T$, termed a 0 region. Thus a time-reversed Andreev pair formed by the usual Andreev retroreflection at the superconductor/graphene interface has, near the CNP, a large probability of encountering a n/0 or p/0 boundary. At such boundaries such junctions, a specular-like reflection must occur when two counter propagating electrons diffusing in the n-doped region are converted into two electrons belonging to two different bands in the 0 region. The change in relative velocity destroys the counter propagation of the pair. As the two electrons diffuse across the rest of the graphene, they undergo uncorrelated scattering events and their relative phase difference increases. Since the total supercurrent is the sum of all contributions from the propagating Andreev pairs, constructive interference is destroyed when counter-propagation is lost, and thus the supercurrent is suppressed (see Figure 6). Interestingly, the effect

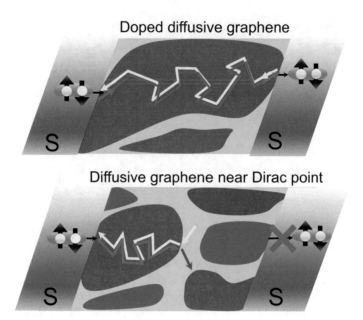

FIGURE 6. Sketch of the superconducting proximity effect through diffusive graphene, at high and low doping. Top sketch: Highly doped regime. The usual Andreev retroreflection at the S/G interface leads to diffusive counterpropagation with zero total phase accumulation. Bottom sketch: Low doping regime. Specular Andreev reflection of propagating Andreev pairs can occur at an n/0 or p/0 junction, leading to loss of counterpropagation and thus large phase accumulation within an Andreev pair. Supercurrent, which results from all Andreev trajectories, is destroyed. The red region is electron doped, the blue one hole doped, and the green region in between has zero doping ($E_F < E_{Th}$).

of these puddles is immense in the superconducting state (and presumably all the more so that the superconducting coherence length, the "size" of the pair, is small with respect to the puddle size), whereas it is much weaker in the normal state where thanks to Klein tunneling, the puddles do not suppress single quasiparticle propagation so much.

In summary, whereas the specular Andreev reflection in ballistic S/G/S junctions can yield a supercurrent [35], we have shown that in diffusive S/G/S junctions a specular-like reflection of Andreev pairs at p/0 or n/0 junctions leads to accumulation of phase difference within the Andreev pair. The critical current is then suppressed, in a manner which depends on the number of such n/0 (or p/0) junctions within the sample. This translates into a critical current suppressed most near the charge neutrality point. The supercurrent suppression by charge puddles is thus expected to be largest in samples that are long (large ratio of sample length

L to puddle size, typically larger than 50 nm [33]) and connected to superconductors with large gaps, corresponding to smaller superconducting coherence lengths ($\xi_s = \sqrt{\hbar D/\Delta}$ is typically 125 nm in graphene for Nb ($\Delta = 1.6$ meV) or 170 nm for ReW ($\Delta = 1.2\ meV$, as compared to 350 nm for Al ($\Delta = 0.2$ meV), given the diffusion constant $D = 4.10^{-2}m^2/s$ in these graphene samples.

4. Perspectives: inducing new functionalities in graphene by creating scattering centers

We have discussed in the previous sections how impurity scattering and charge inhomogeneities affects transport properties of disordered graphene. We will now discuss to what extent it is possible to induce new electronic properties in graphene by functionalisation with scattering centers which can be of very different nature.

4.1. Adatoms, agregates or molecules

It has been already shown that charged impurities give rise to anomalies in the density of states depending on the position of the Fermi energy. When the charge is larger than a critical value these evolve into huge peaks identified as "collapse" states by analogy with the capture of relativistic electrons by large Z positively charged nucleus [37]. It has also been predicted that magnetic adatoms can give rise to gate dependent Kondo screening and exchange coupling with electron or holes carriers in graphene [38, 39]. This gate induced magnetism has been observed with the deposition of molecules on graphene [40]. Another very interesting functionality which can be induced in graphene is spin orbit coupling. It was shown indeed that the band structure of graphene with spin orbit interactions can be completely modified, with the opening of a gap at the Dirac points and the formation of topological spin polarized edge states. The intrinsic spin orbit of carbon atoms being extremely small, it has been proposed to deposit adatoms on graphene. Depending on the strength of their chemical bonding and their positions with respect to the hexagonal lattice these atoms can either induce a so-called "intrinsic" spin orbit or a Rashba interaction described respectively by the Hamiltonians H_{SO} and H_R:

$$H_{SO} = \Delta_{SO}\tau_z\sigma_z s_z$$
$$H_R = \Delta_R(\tau_z\sigma_x s_y - s_x\sigma_y),$$
(5)

where τ_z is the valley index, $\vec{\sigma}$ and $\vec{s}$ are respectively the sublattice and real spins. The most interesting component is H_{SO}, proportional to σ_z therefore breaking the sublattice symmetry, which can turn graphene into a topological insulator. Calculations based on density functional theory [41, 42], have shown that graphene can inherit strong spin-orbit coupling from a dilute concentration of heavy adatoms randomly deposited onto the honeycomb lattice. The atomic SOI of these elements can be of the order of an electron-Volt. The SOI induced in graphene can be understood by considering feels in which an electron from graphene tunnels onto an adatom whereupon it feel the spin-orbit coupling and then returns to the graphene sheet. Depending on the position of the adsorbed atoms with respect to

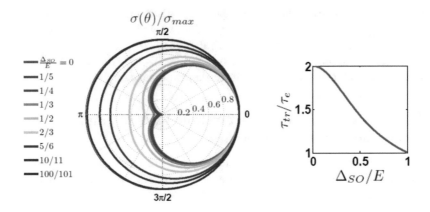

FIGURE 7. Evolution of the angular dependence of the scattering crossection on resonant impurities with increasing spin orbit strength. On can see that an angular dependence is recovered at large spin orbit coupling which indicates that intravalley back scattering is not forbidden anymore. Inset: evolution of the ratio τ_{tr}/τ_e, which varies between 2 to 1 (from [43]).

the hexagonal cell, their concentration, and the microscopic nature of the induced SOI, different behaviors have been predicted, from the generation of a spin Hall effect, to a spin orbit band splitting or a spin Hall insulator with spin filtered edge states. Indium and thallium atoms are expected to occupy sites at the center of the hexagons of the honeycomb lattice, transforming graphene into a spin Hall insulator. On the other hand, certain adatoms like osmium and iridium form spin-orbit-split impurity bands that hybridize with graphene Dirac states. It is then more appropriate to view the adatoms as the dominant low-energy degrees of freedom, with their coupling effectively mediated by graphene. However whereas the configuration of diluted atoms has been investigated in detail, the more realistic situation where atoms form clusters or islands on graphene still needs to be investigated theoretically.

It was recently suggested [43] that measuring the ratio τ_{tr}/τ_e as discussed in Section 1 can provide information on the amplitude of the spin orbit coupling. Asmar and Uloa have shown that SOI leads to an important transformation of short range scattering low energy processes in graphene from highly anisotropic (zero back scattering) to more or fully isotropic, depending on the strength of these interactions. This is shown in Figure 7 showing the evolution of the angular dependence of the scattering cross section and τ_{tr}/τ_e with the strength of Δ_{SO}.

On the experimental side it has turned out that depositing adatoms on graphene in a controlled soft way is not an easy task and can strongly degrade the quality of graphene by the formation of extra defects and clustering.

An alternative approach for inducing spin-orbit coupling in graphene is to functionalize samples with organo-metallic molecules containing heavy non magnetic atoms such as metallo-porphyrins or phtalocyanins. In contrast to evaporated atoms which tend to cluster in a disordered way, these molecules are known to form ordered arrays with a periodicity that can be adjusted by choosing the size of the organic side chains of the molecules. Among them, porphyrins and phtalocyanins are already known to self-organize into well-ordered structures on clean graphite surfaces [44]. When deposited on hole doped graphene, these molecules act as donors and electron-dope the pristine substrate. We have found that the mobility can even be increased [40]. On the other hand hole doping is observed when graphene is initially electron doped. In general we have found that the molecules tend to neutralise charges on graphene. We have chosen to work with metallo-porphyrin molecules with heavy Pt atoms in order to induce large spin spin-orbit interactions. Whereas the amplitude of induced SOI could not be estimated yet, we found, by using superconducting electrodes, that gate-controlled magnetism can be induced in graphene: Pt porphyrins are non-magnetic when they are neutral, but acquire a magnetic moment when they are ionized. We found that the gate voltage controls this ionization process at room temperature, but not at low temperature. It is thus in principle possible to freeze a given ionization state of the porphyrin molecules on graphene by cooling the sample at a chosen gate voltage. The magnetic state of the ionized porphyrins can be subsequently controlled at low temperature as revealed by proximity-induced superconductivity experiments (see Figure 8). These results may be evidence for the long-sought Fermi-level-controlled exchange interaction between localized spins and graphene, leading to new kinds of gate controlled magnetic devices. Measurements under different field orientations exhibit hysteretic behavior, revealing a strong magnetic anisotropy, a promising observation that needs to be understood. For the purpose of demonstrating a spin orbit interaction induced by these molecules, a probable route will be to adjust the doping at room temperature (via the gate voltage) so that the molecules are not ionized and therefore not magnetic. We expect that in this case the induced spin orbit interaction will overcome the magnetic interactions at low temperature.

4.2. Engineering inhomogeneous strain in graphene

In contrast with the modification of the electronic properties of graphene by the deposition of adatoms or molecules, we discuss now another way to engineer new electronic properties in graphene based on the generation of local strained regions. It was pointed out that new non-trivial electronic properties were to be expected in strained graphene nanostructures in which the two atoms in the unit cell experience different potentials. This effect can be described by a polarization of the pseudo spin degrees of freedom under large effective magnetic fields [11]. We have recently found that it is possible to deposit graphene on etched substrates consisting of arrays of silicon oxide nanopillars (see Figure 9). We obtain this way a periodic array of graphene strained bumps separated by ripples. Depending on the lattice

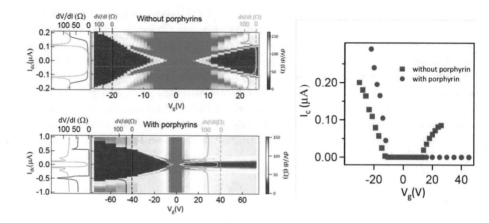

FIGURE 8. Comparison between the proximity effect in graphene connected to Pd/Nb electrodes before (upper) and after (lower) deposition of porphyrin molecules. The color scale codes the differential resistance as a function of gate voltage (x-axis), measured with a small ac current added to the dc current (y-axis). The dark red region corresponds to the region of zero differential resistance where a Josephson supercurrent runs through the S/graphene/S junction. Whereas the Josephson effect occurs symmetrically on both sides of the Dirac point on the bare sample, (bipolar supercurrent) it only occurs on the hole doped side on the sample covered with porphyrins (unipolar supercurrent). The curves on the left of the color plots correspond to the differential resistance curves measured at gate voltages symmetric with respect to the Dirac point. Right: gate voltage dependence of the supercurrent on the bare graphene sample compared to the graphene grafted with porphyins. These findings are interpreted as the signature of the existence of a collective magnetic state for the ionized porphyrins mediated by graphene conduction electrons for positive gate voltages [40].

spacing of the array of pillars, different conformations of graphene were observed ranging from complete suspension of graphene or on the opposite a complete coverage of the substrate with the formation of ripples connecting nearest neighbour pillars. These structures have been investigated in detail in [45] combining electron and atomic force microscopy observations with Raman measurements. Transport properties are under investigation. Both strain inhomogeneities and ripples are expected to strongly affect electronic properties of graphene. Whereas a local strain is expected to induce a polarization of the pseudo spin degrees of freedom acting as an effective magnetic field which modifies the density of states, a ripple scatters charge carriers creating a dephasing field within intravalley transport. In the presence of a gate voltage, ripples can behave as smooth pn junctions leading to anisotropic transport and collimation effects [46]. Moreover graphene bubbles can be considered as quantum dots, the confined energy spectrum of which, has been calculated [47]. Depending on the geometry of the bubble, level crossings are found. These give rise to a strongly modified orbital magnetism under an applied (real) magnetic field expected to be paramagnetic rather than diamagnetic. More spectacular is the possible existence of current loops constituting a magnetic texture around the bubble in the absence of any applied magnetic field [48].

FIGURE 9. Scanning electron microscopy picture of CVD Graphene stretched on a network of 300nm high SiO$_2$ nanopillars [45]. Examples of graphene transferred on pillar arrays. Work done in collaboration between LPS Orsay and Neel groups leading to partially (left) or totally (right) suspended graphene.

5. Conclusion

We have shown that transport properties of graphene in the presence of disorder or intentionally deposited adatoms and molecules, bear the signature of its Dirac perticular spectrum as well as the symmetry properties of the wave functions. We have first considered the case of strong resonant scatterers whose range is intermediate between the carbon-carbon nearest neighbour distance (0.3 nm) in graphene and the Fermi wavelength (few nm) which play the dominant role in the limitation of the mobility of graphene. Despite of being short range, these scatterers do not lead to intervalley scattering and are sensitive to the Berry phase factor of the pseudo-spin component of the wave function. In the absence of spin orbit scattering, they are characterized by a ratio $r = \tau_{tr}/\tau_e$ equal to 2 measured for graphene instead of the value of 1 obtained for a single parabolic spectrum. It was pointed out theoretically that the value $r = 1$ can be recovered for impurities with a large spin orbit scattering potential, suggesting that the investigation of this ratio r can yield the strength of the spin Orbit component of the scattering potential. The case of the bilayer still needs to be understood. In the opposite limit very long range scatterers, the second section of this paper was devoted to the sensitivity of quantum transport to the formation of electron hole puddles in the low doping region. A spectacular experimental consequence is the complete suppression of the supercurrent close to the Dirac point in long graphene-based Josephson junctions.

We also have shown that functionalised graphene with grafted porphyrin molecules leads to new properties such as gate voltage dependent magnetism and potentially spin orbit interactions. The possibility to induce periodic inhomogeneous strain networks is also discussed.

Acknowledgment

We thank all our colleagues who largely contributed to the work presented in this paper: Miguel Monteverde, Claudia Ojeda, Jean Noel Fuchs, Katsuyoshi Komatsu, Dimitri Maslov, Keyan Bennaceur, Christian Glattli, Raphael Weil, Sandrine Autier-Laurent.

References

[1] Castro Neto, A.H., et al.: Rev. Mod. Phys. **81**, 109 (2009).

[2] Wallace, P.R.: Phys. Rev. **71**, 622 (1947).

[3] Shon, N., and Ando, T.: J. Phys. Soc. Jpn **67**, 2421 (1998).

[4] Akkermans, E., and Montambaux, G.: *Mesoscopic Physics with electrons and photons*. Cambridge University Press, (2007).

[5] Coleridge, P.T.: Phys. Rev. B **44**, 3793 (1991).

[6] Ojeda-Aristizabal, C., Monteverde, M., Weil, R., Ferrier, M., Guéron, S., and Bouchiat, H.: Phys. Rev. Lett. **104**, 186802 (2010).

[7] Heersche, H.B., et al.: Nature **56**, 446 (2007).

[8] Du, X., Skachko, I., and Andrei, E.Y.: Phys. Rev. B **77**, 184507 (2008).

[9] Ojeda-Aristizabal, C., Ferrier, M., Guéron, S., and Bouchiat, H.: Phys. Rev. B **79**, 165436 (2009).

[10] Dongchan Jeong, Jae-Hyun Choi, Gil-Ho Lee, Sanghyun Jo, Yong-Joo Doh, and Hu-Jong Lee: Phys. Rev. B **83**, 094503 (2011).

[11] Levy, N., Burke, S.A., Meaker, K.L., Panlasigui, M., Zettl, A., Guinea, F., Castro Neto, A.H., and Crommie, M.F.: Science **329**, 544 (2010).

[12] Aleiner, I.L., and Efetov, K.B.: Phys. Rev. Lett. **97**, 236801 (2006).

[13] Ostrovsky, P.M., Gornyi, I.V., and Mirlin, A.D.: Phys. Rev. B **74**, 235443 (2006).

[14] Nomura, K., MacDonald, A.H.: Phys. Rev. Lett. **96**, 256602 (2006); Ando, T.: J. Phys. Soc. Japan **75**, 074716 (2006).

[15] Adam, S., and das Sarma, S.: Phys. Rev. B **77**,115436 (2008).

[16] Adam, S., Cho, S., Fuhrer, M., and Das Sarma, S.: Phys. Rev. Lett. **101**, 046404 (2008).

[17] Jang, C., et al.: Phys. Rev. Lett. 101, 146805 (2008); Ponomarenko, L.A., et al.: Phys. Rev. Lett. **102**, 206603 (2009).

[18] Couto, N.J.G., Sacepe, B., and Morpurgo, A.F.: Phys. Rev. Lett. **107**, 225501 (2011).

[19] Katsnelson, M.I., and Novoselov, K.S.: Solid State Commun. **143**, 3 (2007); Stauber, T., Peres, N.M.T., and Guinea, F.: Phys. Rev. B **76**, 1120 (2007).

[20] Cho, S., and Fuhrer, M.S.: Phys. Rev. B **77**, 081402 (2008).

[21] Abanin, D.A., and Levitov, L.S.: Phys. Rev. B **78**, 035416 (2008).

[22] Williams, J.R., et al.: Phys. Rev. B **80**, 045408 (2009).

[23] Lifshitz, I.M., and Kosevich, A.M.: Sov. Phys. JETP **2**, 636 (1956); Coleridge, P.T., Stoner, R., and Fletcher, R.: Phys. Rev. B **39**, 195412 (1989).

[24] Monteverde, M., Ojeda-Aristizabal, C., Weil, R., Bennaceur, K., Ferrier, M., Guéron, S., Glattli, C., Bouchiat, H., Fuchs, J.N., and Maslov, D.L.: Phys. Rev. Lett. **104**, 126801 (2010).

[25] Hwang, E.H., and Das Sarma, S.: Phys. Rev. B **77**, 195412 (2008).

[26] Adhikari, S.K.: Am. J. Phys. **54**, 362 (1986).

[27] Novikov, D.S.: Phys. Rev. B **76**, 245435 (2007).

[28] Basko, D.: Phys. Rev. B **78**, 115432 (2008).

[29] Meyer, J.C., et al.: Nano Lett. **8**, 3582 (2008).

[30] Katsnelson, M.I., Vozmediano, M.A., and Guinea, F.: Phys. Rev. B **77**, 075422 (2008).

[31] Couto, N.J.G., et al.: Physical Review X 4, 041019 (2014).

[32] Tikhonenko, F.V., Horsell, D.W., Gorbachev, R.V., Savchenko, A.K.: Phys. Rev. Lett. **100**, 056802 (2008).

[33] Martin, J., et al.: Nat. Phys. **4**, 144 (2008).

[34] Beenakker, C.W.J.: Rev. Mod. Phys. **80**, 1337 (2008).

[35] Titov, M., and Beenakker, C.W. J.: Phys. Rev. B **74**, 041401(R) (2006).

[36] Katsuyoshi, Komatsu Chuan, Li, Autier-Laurent, S., Bouchiat, H., and Guéron, S.: Phys. Rev. B **88**, 115412 (2012).

[37] Yang Wang, et al.: Science **340**, 734 (2013).

[38] Uchoa, B., Kotov, Valeri N., Peres, N.M.R., and Castro Neto, A.H.: Phys. Rev. Lett. **101**, 026805 (2008).

[39] Kotov, Valeri N., Uchoa, B., Pereira, Vitor M., Castro Neto, A.H., et al.: Rev. Mod. Phys. **84**,1067 (2012).

[40] Chuan Li, et al.: Phys. Rev. B 93, 045403 (2016).

[41] Weeks, C., Hu, J., Alicea, J., Franz, M., and Wu, R.: Phys. Rev. X **1**, 021001 (2011).

[42] Hu, J., Alicea, J., Wu, R., and Franz, M.: Phys. Rev. Lett. **109**, 266801 (2012).

[43] Asmar, Mahmoud M., Ulloa, Sergio E.: Phys. Rev. Lett. **112**, 136602 (2014).

[44] Otsuki, J., Kawaguchi, S., Yamakawa, T., Asakawa, M., and Miyake, K.: Langmuir **22**, 5708–5715 (2006).

[45] Reserbat-Plantey, A., et al.: Nanoletters **14**, 5044 (2014).

[46] Cheol-Hwan Park, Young-Woo Son, Li Yang, Cohen, Marvin L., Louie, Steven G.: Nano Lett. **8**, 2920–2924 (2008).

[47] Kyung-Joong Kim, Blanter, Ya.M., and Kang-Hun Ahn: Phys. Rev. B **84**, 081401 (2011).

[48] Guinea, F., Katsnelson, M.I., and Geim, A.K.: Nature Physics **6**, 30–33 (2010); Wakker, G.M.M., et al.: Phys. Rev. B **84**, 195427 (2011).

Chuan Li, Sophie Guéron and Hélène Bouchiat
LPS – Bât. 510, Université Paris Sud, CNRS, UMR 8502
F-91405 Orsay cedex, France
e-mail: chuan.li@u-psud.fr
       sophie.gueron@u-psud.fr
       helene.bouchiat@u-psud.fr

Dirac Matter, 75–94
© 2016 Springer Basel AG

Experimental Signatures of Topological Insulators

Laurent Lévy

Abstract. Energy bands in solids describe quantum states in periodic crystals. When a quantum state is wound around the Brillouin zone, it acquires a quantum phase. For a completely filled band, the global phase acquired in this winding is a topological property of the band. For 3D solid with time-reversal symmetry, there are two topological classes corresponding to a ± 1 sign. A signature of topological solids (minus sign) is the presence of conducting surface states with a relativistic dispersion, similar to graphene. They can be observed in angle resolved photo-emission which is able to reconstruct their energy-momentum dispersion below the Fermi level. Some of the experimental signatures of these topological states in strained Mercury-Telluride are presented: their Dirac spectra measured at the SOLEIL synchrotron, the ambipolar sign of their surface charge carriers, the topological phase in their Landau-level quantization, and the weak-antilocalization peak in magneto-transport also controlled by the π-topological phase.

1. Introduction

In condensed-matter physics, the quantum mechanical phase is often considered as a fragile quantity subject to thermal fluctuations, disorder or even structural defects. The usual cases when the phase plays a role in macroscopic samples is when the carriers condense in a macroscropic ground state, with the opening of an energy gap which offers a protection against thermal fluctuations.

This is what happens in superconductors, where the electrons condense in a macroscopic BCS quantum state which has a lower energy than the usual electron-Fermi liquid. The phase of the ground state is a macroscopic quantity which is robust and rigid. This phase rigidity against electromagnetic perturbations leads to the Meissner effect expelling a magnetic field from the bulk material.

In low-dimensional systems, fundamental symmetries impose definite properties to the quantum mechanical phase, and the usual dephasing processes can be forbidden on symmetry grounds. A classic example is the difference between gap-less behavior of half-integer spin-chains which contrast with the gapped structure of integer spin chains, the difference coming from the 4π periodicity of half-integer spin wavefunctions compared to the 2π periodicity of integer spins. The sign difference under a 2π rotation may seem harmless but it kills most quantum fluctuations as the corresponding quantum tunnelling processes become forbidden. The boundaries (end of the chain) define places in space where the symmetries change. This usually leads to special excitations at the boundaries which are one of the manifestations of a change in the fundamental symmetries.

Topological insulators have stirred a great deal of interest, because there is a symmetry in band structure theory which had not been considered before which leads to similar properties as the one encountered in one-dimensional magnets. The key difference is of course that the symmetry induced by the properties of the quantum phase become truly macroscopic and can indeed be observed in two (quantum spin-Hall effect) and three dimensions (Dirac matter at surfaces).

The problem of topological band structure in solid is illustrated in Figure 1, [1, 2] where we compare the electronic wave functions and the band structure of solid Neon or Argon, which are atomic insulators with the one of 2D electronic gases in large field. For atomic insulators, the possibilities of hopping between sites are limited by the large splitting between filled atomic shells. Ignoring hopping, all electrons in the filled band have the same shell energies and no low-energy excitations are possible: these solids are indeed strong band insulators with no electrical conductivity. This situation is reminiscent of what happens in a 2D electron gas, where all the electrons condense in Landau level where they have all the same energies. Provided that the chemical potential lies between Landau levels, these macroscopic states are filled and no low energy excitations are available: indeed the bulk longitudinal conductivity is zero. However, in 2D quantum Hall systems, there are n-edge states forming at the boundaries which signals that there is a fundamental symmetry change between the interior of the 2D system and vacuum: the consequence is a finite Hall conductivity $\sigma_{xy} = n\frac{e^2}{h}$, directly related with the number of edge states at the boundary. The index n is in fact the difference between two topological indices, the index of the 2D quantum Hall state (n) and the one of vacuum (0).

This difference between atomic insulators and quantum-Hall states shows that all the information is not simply contained in the energy spectrum, but that the ground state wavefunction and more specifically its phase matters and leads to macroscopic differences, here in the Hall conductivity σ_{xy}.

The notion of Brillouin zone (or magnetic Brillouin zone for quantum Hall systems) is essential to understanding how topological quantum numbers may arise in solids. There is a natural atomic periodicity in 3D crystalline solids, which allows to identify quantum states with a wavevector q_x in the $\hat{x}$ direction with a state with wavevector $q_x + m\frac{2\pi}{a}$ where a is the lattice periodicity in the x direction. This means that only states with wave-vectors in the range $[-\frac{\pi}{a}, \frac{\pi}{a}]$ need to be

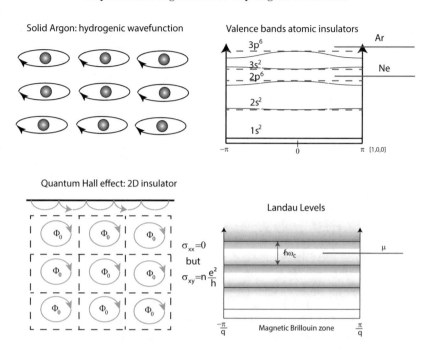

FIGURE 1. Different solids, may have the same band spectra, but different electronic properties. In an atomic solid, such as solid neon or solid
argon, atomic shells are filled and the corresponding bands have little
dispersion. The chemical potential of such atomic solid lies in between
filled and empty bands and there is no electronic transport possible as
thermal excitation cannot excite electrons across the shell gaps. A two-
dimensional electron gas in a magnetic field has a similar spectrum, and
when the chemical potential lies in between Landau levels, the longitudinal conductivity is zero. In this quantum Hall regime, the system has
a quantized Hall conductivity in units of $\frac{e^2}{h}$, where n is the number of
edge states. The presence of gapless excitations at the boundaries is a
signature of a change of symmetry, here a topological symmetry class.

considered. Since the states at $-\pi/a$ and π/a are identical, the Brillouin zone has
the topology of a torus. Suppose now that by applying an electric field for a finite
time in this same direction $\hat{x}$, we take a quantum state from q_x to $q_x + \frac{2\pi}{a}$, i.e.,
one makes a closed loop around one of the Brillouin zone section of the torus.
The final state being the "same" as the initial state, it can only differ by a phase
factor after this winding around the Brillouin zone section. In general this phase
depends on the specific way, it was wound across the Brillouin zone. However if we
repeat the operation for the wavefunction of all the electrons in a filled band, the
overall phase acquired by this wavefunction can only take two values 2π or π for a
3D solid with time-reversal symmetry. It is in fact time reversal symmetry which

imposes this phase factor to be real, leaving ±1 as the only possible values. This amounts to a sign change in the filled band wavefunction when winding around the Brillouin zone. Since there are usually several filled band in a 3D solid, the signs corresponding to all the filled bands have to be compounded. This gives one of the procedure defining the topological index of a band-insulator: since it is just sign the overall symmetry is Z_2. When the sign is $+1$, the material is an ordinary insulator, while the -1 sign describe the new class of topological insulating materials.[1]

When the spin-orbit interaction is weak, Fu and Kane [3] showed that the topological index C of a band insulator depends on the number p of P bands below the Fermi level: $C = (-1)^p$. This relation offers a search route for real material that are topological insulators. Any material class where a band inversion between S and P bands takes place will change the value of p by one and turn a band insulator into a topological insulator. There are several materials with a so-called negative gap, i.e., with an inverted band structure. Among them, we find Cadmium Telluride (a normal insulator) with a gap $\Delta = 1.51\text{eV}$ between the Γ_6 (S band) and the Γ_8 (P band) and its topological counterpart Mercury Telluride where this gap is negative $\Delta = -0.3$ eV (inverted bands). Other pairs of materials with gaps of opposite signs and topologies are $\text{Bi}_{1-x}\text{Sb}_x$ and BiSb.

We now consider two materials with different topological symmetries. The filled band of a topological insulator cannot be connected to an insulator filled band, because they differ by their topological index. Hence as the interface between the two materials is crossed, the only possible correspondence is between a filled valence band and an empty conduction band. Another equivalent argument is that the connection between bands can only involve the same orbital symmetries: the Γ_8 band (P) of HgTe above the Fermi level connects with the filled Γ_8 band in CdTe. This is represented schematically on Figure 2.

We see that the bands cross at the surface, leading to a metallic metallic surface states. The analogy with edge states in the quantum Hall effects which signal a change in topological class, these surface states signal a change of topological symmetries at the interface and are considered as the hallmark of a material with a nontrivial band topology. Because of Kramer's theorem, the band crossing takes place at one of the symmetry points of the Brillouin zone, usually the zone center. As one moves away from this point into the Brillouin zone, spin-orbit contributions split the Kramer pair linearly, giving a linear dispersion in crystal momentum k. The presence of a Dirac cone is one of the very distinctive property of topological surface states which can be contrasted with other massive (Shockley [4] and Tamm [5]) surface states found in condensed matter physics. The states on the Dirac cone emerging from a Kramer doublet are timed reversed states: this gives a helical spin structure (the spin of surface charge carrier are tangent to the cone, at least close to the Dirac point). Symmetry arguments do not specify the Dirac point energy, besides the fact that it is between the inverted bands. This complicates the

---

[1] For three-dimensional insulators, the accumulated phases are additive when wrapping the Brillouin zone torus on a sphere: the topological index in an integer, with topological symmetry Z.

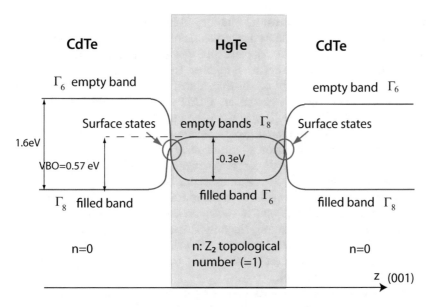

FIGURE 2. A thick slab of Mercury Telluride is sandwiched between a Cadmium Telluride substrate and cap-layer. The material growth direction is $\hat{z}$. We consider the center of the Brillouin zone and represent the extrema of the Γ_6 (S band $J = 1/2$) and the $\Gamma_{8,\mathrm{LH}}$ ($J = 3/2, m_J = \pm 1/2$). For simplicity, the heavy hole band is not represented. The filled $\Gamma_{8,\mathrm{LH}}$ of CdTe and the filled Γ_6 band of HgTe have different topological symmetries and cannot be connected across the interface. One the other hand the filled CdTe $\Gamma_{8,\mathrm{LH}}$ band has connects to the empty HgTe Γ_6 band with the same orbital symmetry. The same thing happens between the filled HgTe Γ_6 band and empty CdTe $\Gamma_{8,\mathrm{LH}}$ band. These band necessarily cross at the interface, where a gapless surface state appears as a result of the inverted symmetries.

picture significantly because the 3D bulk bands are also present and the interaction between bulk and surface bands can lead to hybridized states. Furthermore, in electronic transport, a 'pure' conductivity response coming from the surface states alone are hard to observe, because spurious conduction involving bulk states often contaminates the measured electrical conductivity tensor (σ_{xx} and σ_{xy}).

2. Bulked gap in strained mercury telluride

The band structure of Mercury Telluride is shown on Figure 6. The Γ_8 band being a $P_{3/2}$ band, it is split by spin orbit into $\pm 3/2$ (Γ_{8hh}) (heavy holes) and $\pm 1/2$ (Γ_{8lh}) (light hole) components. The Γ_{8lh} is the conduction band (positive charge carriers) while the Γ_{8hh} is a valence band. In absence of strain, they are degenerate at the zone center (Γ-point). As a result, Mercury-Telluride is a semi-metal with bulk-bands present everywhere in the energy spectrum. The Γ_6 band being 0.3 eV below the Γ_{8lh}, topological surface states can still develop in between, but the coexistence of bulk and surface state is an issue at least in transport experiments. For this reason, it was suggested early on [6] that a strain-gap between the two Γ_8 bands could be open when HgTe is grown on a CdTe substrate which lattice constant ($a_{CdTe} = 6.48$Å is 0.3% times larger than the one in HgTe ($a_{HgTe} = 6.46$Å). This small lattice mismatch allows to grow homogeneously expanded HgTe layers up to 150–250 nm before dislocation spontaneously appears. At such thicknesses, the material is well in the three-dimensional regime. As will be shown in the next section the crossover to the 2D regime occurs between 15 and 25 nm. The 3D-spectrum of the strained layers can be computed using the Kane-model [7] with the Bir–Pikus Hamiltonian [8] and is shown on the left panel of Figure 6. Although the strain-gap magnitude is small, the only states which are present in this energy interval are the topological surface states (black lines). The only unknown is the actual position of the Dirac point with respect to the top of the Γ_{8hh} valence band and the band velocity (slope).

This information is best obtained using ARPES experiments which measures the energy spectrum of the occupied states in a solid. This technique is particularly suited for probing surface states as the photons penetrate 1nm into the sample, less than the surface states width.

3. ARPES spectra and surface mercury telluride

The photon energy used in ARPES experiments is usually in the 15–30 eV range, above the material workfunction ($\Phi = 5.8$ eV for HgTe) but below the energy of Mercury and Tellurium core levels. In this range, the relationship between the measured kinetic energy E_{kin} of the extracted electron and the binding energy $E_B(k)$ of its original occupied state is [9]

$$h\nu = E_{kin}(\vec{K}) + \Phi + E_B(\vec{k}), \tag{1}$$

$$\frac{h\nu}{c}\hat{P} = \vec{K} - \vec{k}; \tag{2}$$

$\hat{P}$ is the direction of the incident photon, $\vec{K}$ is the momentum wavevector of the extracted electron in vacuum ($E_{kin} = \frac{(\hbar K)^2}{m}$) while $\vec{k}$ is the crystal wavevector of the electron in the solid. ARPES is a useful technique only when the surfaces are sufficiently clean so that a definite relationship between the crystal wavevector and the (measured) wavevector of the extracted electron can be established. For a pristine surface,

$$k_\parallel = K_\parallel = \frac{\sqrt{2mE_{kin}}}{\hbar}\sin\theta, \tag{3}$$

$$k_\perp = \frac{\sqrt{2m(E_{kin}\cos^2\theta + V_0)}}{\hbar}, \tag{4}$$

where the crystal wavevector $k_\parallel$ parallel to the surface, preserved as the electron crosses the surface, has to be understood in an extended zone scheme (i.e., with the possible addition of a reciprocal lattice wavevector). Here $V_0 = \Phi + E_0$ which is the sum of the workfunction and the energy of the bottom of the band is the "inner" potential. On Figure 3, we have represented the electron momenta on each side of the interface. For a bulk bands, there is a range of perpendicular momenta which contribute to the photoelectron intensity which is determined by the band width. When the photon energy is decreased so do the radii of accessible momenta and the range of perpendicular momenta contributing to the photoelectron intensity. For surface electrons, there is only the momentum which matches $K_\parallel$ which contribute to a given photo-electron direction ($k_\perp$ is undefined because of

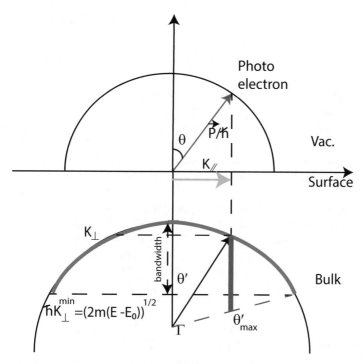

FIGURE 3. Top part: wavevector of the outgoing photo-electron and its projection, which is conserved for a clean interface. For a surface state, it is the electron wave-vector (green arrow). Bottom, the surface with constant energy is represented for bulk states. There is a range of $k_\perp$ momenta perpendicular to the surface which satisfy the conservation of energy and the transverse momentum. As the photon energy is changed this range will vary, changing the intensities maps of the photoemitted spectrum.

the interface discontinuity), irrespective of the photon energy used. The dispersion $E_B(K_\parallel)$ gives the surface band dispersion without any further assumption.

The experimental ARPES spectra of homogeneously strained 100 nm thick HgTe slabs epitaxially grown on [100] CdTe substrates were obtained on the CASSIOPEE line at the SOLEIL synchrotron, whose low energy photons and high resolution (few meV) spectrometer are well suited to topological insulator studies. The strained HgTe slabs were grown by low temperature Molecular Beam Epitaxy in a Ultra-High Vacuum chamber from a [100] CdTe substrate [10]. Only occupied electronic states can be observed in ARPES: indium-doped samples at 10^{18}cm^{-3} were also prepared in addition to the un-doped reference samples, in order to raise the bulk chemical potential. The samples surfaces which spontaneously oxidize in air, were cleaned in a dedicated Ultra High Vacuum preparation chamber by a low-energy Ar-ion sputtering at grazing angles to remove the surface oxide. The in-situ LEED spectra observed after surface cleaning (Figure 5) shows a c(2x2) reconstructed pattern consistent with an HgTe [100] growth. The samples were subsequently transferred to the ARPES chamber in Ultra-High Vacuum. The position of the Fermi level was determined with a reference gold sample placed on the same sample holder.

The high-resolution spectra in the vicinity of the Γ-point for an un-doped sample are shown in Figure 4 [11]. On the left panel a), the intensity of the ARPES spectrum is shown for an incident photon energy $h\nu = 20$ eV. The surface projection of the two volume valence bands $\Gamma_{8,HH}$ and Γ_6 (deep blue) are observed and, with more intensity (darker), a linear cone structure, which broadens as one moves away from its apex. The second derivative spectrum shown on panel b) enhances the contrast in the ARPES intensity. Within the experimental accuracy the cone apex coincides with the top of the $\Gamma_{8,HH}$ band and lies 0.1 eV below the Fermi level. On the raw ARPES spectrum shown in panel a) the cone structure extends in the gap with a decreasing intensity, as those states are populated mostly through the room-temperature thermal activation. The cone section for different binding energies shown on the bottom panel are circular up to energies 0.4 eV below the Dirac point. In comparison the heavy hole band $\Gamma_{8,HH}$ is strongly anisotropic in the $k_x - k_y$ plane.

From the experimental slope of the cone structure, the surface state band velocity is found to be $v_F \approx 5 \times 10^5$m.s^{-1}. This value agrees with the lowest-order expansion for the energy close to the Dirac point in the Kane model ($\hbar v_F \approx \alpha \frac{P}{\sqrt{6}}$), where the parameter $\alpha \approx 0.9$ for HgTe (the Kane parameters are defined in Ref. [11]) . The same sample was also probed at different incident photon energies $h\nu$. We have seen how different incident photon energy shifts the binding energy of bulk bands according to their k_z dispersion. Here, the cone position is unaffected, emphasizing that this cone structure comes from a surface state with no k_z dispersion (see Figure 5 in [11]). As explained earlier, this is a powerful check which discriminates between 2D and 3D states. Surface state spectra were also collected over the entire Brillouin zone. In the Γ-K direction, the surface state

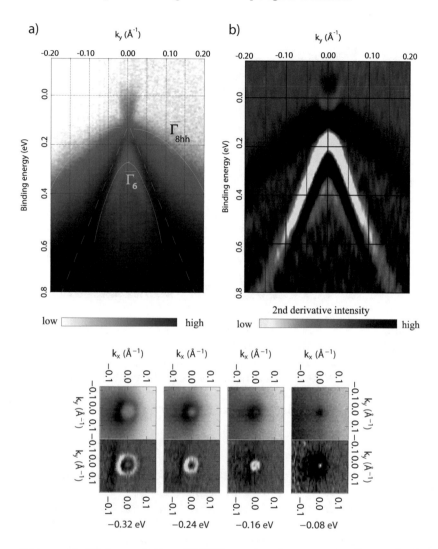

FIGURE 4. High resolution ARPES spectra for a maximally strained [100] HgTe/vacuum interface in the vicinity of the Γ-point measured at room temperature. a) Energy-momentum intensity spectrum after background substraction. b) The second derivative of the intensity data enhances the contrast, although band positions are less faithful. Bottom: Intensity spectrum at different energies. The cone structure has a circular section up to ≈ 0.4 eV.

spectrum becomes diffuse at energies of 0.8 eV below the Fermi level. On the other hand, in the Γ-X direction, the surface state spectra remain linear all the way to the X point (see Figure 5), where its energy is 3.4 eV below the Dirac point, i.e.,

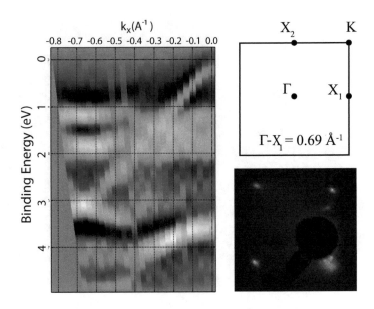

FIGURE 5. Left: Γ to $X1$ ARPES spectrum. The Dirac spectrum observed close to the Γ-point extends all the way to the X point with the same linear slope. Top-right: symmetry points of the Surface Brillouin zone. Bottom-right: LEED pattern observed after low energy Ar^+ surface cleaning: the observed c(2x2) pattern reconstructed surface is in agreement with the half monolayer structures observed on HgTe [100] surfaces.

well below the Γ_6 band: in this direction, the surface state robustness goes well beyond the usual topological protection arguments.

This experimental data can be reproduced in great details using the successful Kane model [7], for which all the parameters [12] are known for HgTe. Since we are interested in an inhomogeneous situation where a topological insulator is in contact with a band insulator or vacuum, we discretized the Kane model perpendicular to the interface between an HgTe slab and vacuum with the 6 bands $(\Gamma_{6,\pm1/2}, \Gamma_{8,\pm3/2,\pm1/2})$ Kane model. The results are shown to be independent of the discretization constant a over the range of energy and momenta considered. The results are shown in Figure 6: on the left panel the energy spectrum of all states are represented. The surface states, originating from the inversion between the two $S = \frac{1}{2}$ bands Γ_6 and $\Gamma_{8,LH}$, are the only states present in the gap. For the Kane parameters used, the energy of the Dirac point is $\epsilon_D = -30\,\text{meV}$ below the $\Gamma_{8,LH}$ and is similar for a CdTe/HgTe interface. At $k = 0$ (ϵ_D), the surface states do not couple to the $\Gamma_{8,HH}$ band, and are weakly affected by the $\Gamma_{8,HH}$ band at small k below ϵ_D. The surface character of the linear spectrum can be verified by

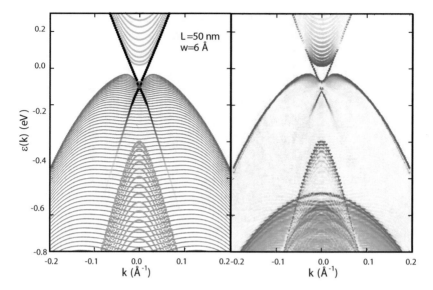

FIGURE 6. Left: energy spectrum obtained with a inhomegeneous discretized Kane model One observes in addition to the bulk $\Gamma_{8,\text{LH}}$ (orange), $\Gamma_{8,\text{HH}}$ (magenta), Γ_6 (green) subbands, a linear dispersive band (black) with a Dirac point 30meV below the top of the $\Gamma_{8,\text{HH}}$ valence band. Right: the color-coded electron density at the surface integrated over a 5 nm depth: the broadening observed in the experiment is reproduced by a decrease in surface electron density.

projecting the overall surface density over a 5 nm thickness, which is the extension of the surface states at the Dirac point. Their dispersion is linear with the same band velocity as in the experiment. Half of the Dirac cone lies inside the $\Gamma_{8,\text{HH}}$ valence band while the other half continues in the stress gap. When the density is integrated over 1.2 nm, the penetration of the synchrotron radiation, the intensity in the gap decrease, as expected since the penetration of surface states is 5 times larger, a detail which is also observed in the experimental data.

The surface state intensity disappear gradually for larger k, consistent with the observed broadening in the experiment. Their dispersion is linear with the same band velocity as in the experiment.

A hallmark of topological insulators is the helical spin structure of surface states induced by the strong spin-orbit coupling. Such textures have been observed directly using spin-resolved ARPES [13, 14, 15]. This helical spin texture around the surface-states Dirac cone also induces circular dichroism in ARPES, which offers another way to probe this helical spin-texture [16, 17]. Circular dichroism is defined as the asymmetry between the ARPES intensity for left (L) and right (R) circular polarization

$$\mathcal{C}(\epsilon, k) = \frac{I_R(\epsilon, k) - I_L(\epsilon, k)}{I_R(\epsilon, k) + I_L(\epsilon, k)}. \tag{5}$$

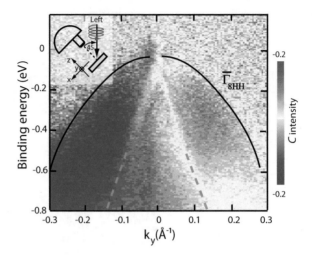

FIGURE 7. Circular dichroism measured at $k_x = 0\text{Å}^{-1}$. The largest contribution to the dichroism comes from the $\Gamma_{8,\text{HH}}$ valence volume band. The surface states appear here as white lines (no dichroism). The inset shows an incident left polarized light beam at $45°$ with respect to the sample surface.

It is plotted in Figure 7 (the geometry is specified in the inset) as a function of k_y for $k_x = 0$ and an incident light beam at $\approx 45°$ with respect to the normal to the sample. By symmetry, the circular dichroism must cancel in the $k_y = 0$ plane as observed for incident photon energies $E_{\text{kin}} > 15.9\,\text{eV}$.

The most salient features of Figure 7 are (i) the absence of dichroism from the surface states, signaled by the white lines (no dichroism) along the surface states dispersion and (ii) a significant dichroism (up to 20%) is observed in the bulk $\Gamma_{8,\text{HH}}$ band. These results on strained HgTe differ from the circular dichroism ARPES data on Bi_2Se_3 compound [16, 17] where a dominant signature of surface states was observed. The relationship between the circular dichroism and the spectral spin densities of low energy bands is complex and depends on the incident photon energies [18]. On the other hand if we assume that such a relationship exists, Wang et al. [16] have shown that the dependence of the ARPES polarization asymmetry on the band polarizations, $\langle S_x \rangle$ and $\langle S_z \rangle$ is

$$\mathcal{C}(\epsilon, k, \phi) = -a^2 \cos\phi \langle S_z(\epsilon, k)\rangle + 4ab \sin\phi \langle S_x(\epsilon, k)\rangle, \qquad (6)$$

for a circularly polarized light beam incident in the x-z plane at an angle ϕ with respect to the normal to the sample. The matrix elements a and b depend on surface symmetries of the material [16]. This formula is consistent with the experimental data of Figure 7 if the coefficient b vanishes for the [100] HgTe surface. This explains the weak circular dichroism contribution of the surface states, whose spin polarization normal to the surface $\langle S_z(\epsilon, k)\rangle$ vanishes at low energy. An analysis if the circular dichroism selection rules in ARPES appropriate for the square surface lattice symmetry, confirms that b must vanish at small k. The direct observation of surface states of stressed Mercury Telluride, confirm its topological insulating nature. It has some quite unique features: the Dirac point sits at the top of the heavy hole band. We now turn to their experimental signature in transport experiments.

4. Topological signatures in transport experiments

As seen in the introduction, the Bloch states for the band α in a periodic lattice $|\psi_\alpha(\vec{k})\rangle = e^{i\vec{k}\cdot\vec{R}}u_\alpha(\vec{k})|\vec{R}\rangle$ are phase sensitive. One can construct wavepackets from them and study their semi-classical motion in electric and magnetic field. Topological effects arise from the phase accumulated in a variation of the wavevector $\vec{k}$ in the course of its motion. Just like the Aharonov–Bohm phase, this phase is a line integral of a "Berry" vector potential, $\vec{\mathcal{A}}_\alpha(\vec{k}) = i\langle u_\alpha(\vec{k})|\vec{\nabla}_{\vec{k}}u_\alpha(\vec{k})\rangle$. When integrated over a loop $\mathcal{C}$ in $\vec{k}$ within the Brillouin zone, this is an Aharonov–Bohm like phase $\Gamma_\mathcal{C} = \oint_\mathcal{C} \mathcal{A}_\alpha(\vec{k})$. Depending on the experimental context, such phase factor are observable in experiments. In a low field magneto-conductance experiment, $\Gamma_\mathcal{C}$ affects interference terms at the origin of the localization peak/dip around zero field. In large magnetic field, the motion of charge carriers gets quantized, and this phase affects the quantization condition. In other transport experiments, this term can be incorporated in the semiclassical equations of motion in the form of an effective magnetic field $\vec{F}(\vec{k}) = \vec{\nabla}_{\vec{k}} \times \vec{\mathcal{A}}_\alpha(\vec{k})$ (Berry's curvature) acting on the momentum exactly as a magnetic field does on the position. Explicitly, we define wavepackets from Bloch eigenstates $\psi_{\vec{k}}^\alpha(\vec{r}) = e^{i\vec{k}\cdot\vec{R}}u_{\vec{k}}^\alpha(\vec{r})$, and compute their center position with the operator $\vec{r} = \frac{\vec{\nabla}_{\vec{k}}}{i} + \vec{\mathcal{A}}(\vec{k})$. The last term introduce a "the shift in the position" which modifies their Poisson brackets $[r_a, r_b] = i\epsilon_{a,b,c}F_c(\vec{k})$. As shown by Haldane [19] and Fuchs [20], this modifies the semiclassical equation of motion,

$$\hbar\frac{d\vec{k}}{dt} = e\left(\vec{E} + \frac{d\vec{r}}{dt} \times \vec{B}\right), \tag{7}$$

$$\hbar\frac{d\vec{r}}{dt} = \vec{\nabla}\epsilon_\alpha(\vec{k}) + \hbar\frac{d\vec{k}}{dt} \times \vec{F}(\vec{k}), \tag{8}$$

through the last term in the motion for the position $\vec{r}$. The Berry curvature $\vec{F}(\vec{k})$ acts in k-space exactly as a magnetic field does in real space. This $\vec{r}-\vec{k}$ duality can be made more explicit by defining the matrix $J(\vec{k})_{ab}$ as the 2×2 Poisson bracket matrix,

$$J(k)_{ab} = -i\begin{pmatrix} [k_a, k_b] & [k_a, r_b] \\ [r_a, q_b] & [r_a, r_b] \end{pmatrix} = \begin{pmatrix} \frac{e}{\hbar}\epsilon_{abc}B^c & -\delta_a^b \\ \delta_b^a & \epsilon_{abc}F^c(\vec{k}) \end{pmatrix}, \tag{9}$$

and the semi-classical equations of motion take a perfectly dual form

$$\hbar J(\vec{k})\frac{d}{dt}\begin{pmatrix} r^b \\ q^b \end{pmatrix} = \begin{pmatrix} \nabla_a H\left(\vec{r},\vec{k}\right) \\ \nabla_{k_a} H\left(\vec{r},\vec{k}\right) \end{pmatrix} = \begin{pmatrix} \nabla_a U(\vec{r}) \\ \nabla_{k_a}\epsilon_\alpha(\vec{k}) + U(\vec{r}) \end{pmatrix}, \quad (10)$$

where $H\left(\vec{r},\vec{k}\right) = \epsilon_\alpha(\vec{k}) + U(\vec{r})$.

Let us first explore the effect of the Berry flux term $\vec{F}(\vec{k})$ on the quantization condition for a cyclotron orbit in a magnetic field. For this we need to compute the action $\mathcal{S}$ over the closed circular orbit $\mathcal{C}$

$$\mathcal{S}(\mathcal{C})eB + \hbar\oint_{\mathcal{C}} \vec{A}(\vec{k})\cdot d\ell - \pi\hbar = nh. \quad (11)$$

Defining $\Gamma(\mathcal{C}) = \frac{1}{2\pi}\oint_{\mathcal{C}} \vec{A}(\vec{k})\cdot d\ell$ as Berry's phase orbit, we find

$$\mathcal{S}(\mathcal{C})\ell_B^2 = n + \frac{1}{2} - \Gamma(\mathcal{C}). \quad (12)$$

For Dirac carriers $\Gamma(\mathcal{C}) = \pm\frac{1}{2}$, depending on the value of the energy with respect to the Dirac point.[2]

This phase $\Gamma(\mathcal{C})$ can be measured directly using Shubnikov–de Haas oscillations. The resistance maxima in Shubnikov–de Haas oscillations occurs when the chemical potential lies between Landau levels which are full or empty, the broadening coming from disorder or other sources, i.e., they match the minima in the density of states. Gusynin and Shaparov have shown that the conductance oscillation of Dirac-like systems had the form

$$\rho_{xx} = \rho_0(B,T)\cos 2\pi\left(\frac{B_f(\mu)}{B} - \frac{1}{2} + \Gamma\right), \quad (13)$$

where $\rho_0(B,T) = \rho_0 R_D(B,\mu)R(T,\mu)$ includes disorder (Dingle) and thermal broadening factors. The magnetic field frequency $B_f(\mu)$, depends on the value of chemical μ which in the experimental below can be adjusted with an electrostatic gate. The minima of the cosine is reached for values the B_n of B which satisfy

$$2\pi\left(\frac{B_f(\mu)}{B_n} - \frac{1}{2} + \Gamma\right) = 2\pi\left(n - \frac{1}{2}\right), \quad (14)$$

$$\frac{B_f(\mu)}{B_n} = n - \Gamma. \quad (15)$$

In other words for all the values of the chemical potential (gate voltage), the intercept of the minima $\frac{1}{B_n}$ as a function of n gives the value of Γ. On Figure 8, the Shubnikov–de Haas oscillations are plotted as a function of magnetic inverse field. The minima in the Shubnikov–de Haas oscillations are obtained from the

---

[2]One can show that if a monopole of unit charge is placed at the Dirac point, which is a singularity in the phase of the wavefunctions, $\Gamma(\mathcal{C})$ is the flux of this monopole through $\mathcal{C}$. Since $\mathcal{C}$ defines a plane dividing the space in two, only half the topological flux is enclosed in $\mathcal{C}$.

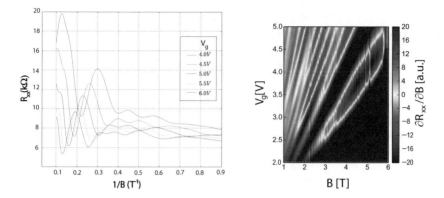

FIGURE 8. Left: Shubnikov–de Haas oscillations of a gated 100nm thick HgTe sample. As a function of gate voltage, the oscillations period changes. Right: the minima in the Shubnikov–de Haas can be located accurately in the Vg-B plane using a color map of the derivative of the longitudinal resistance with respect to the magnetic field. The red and blue zones correspond to a positive and negative slopes, the white line being the extrema. All these extrema converge toward the same gate voltage $V_g \approx 1.3$V, which we interpret as the chemical potential at the Dirac point.

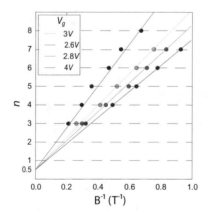

FIGURE 9. Left: Minima $1/B_n$ in the Shubnikov–de Haas oscillations as a function of the Landau index n. For each gate voltage, the data points lie on a straight line which slope is $B_f(\mu)$ and intercept at 0 is $\Gamma \equiv \frac{1}{2}$.

color map shown on the right panel of Figure 8 and re-plotted as a function of the Landau index in Figure 9. For all gate voltage the straight lines converge to the same intercept on the Landau-index axis, which is $\Gamma = \frac{1}{2}$. We see that the topological phase $2\pi\Gamma$ is non-zero and coincide with the π-value expected for Dirac carriers.

The Hall conductivity can be computed from the semiclassical equations of motion [19]

$$\sigma_{ab} = \frac{e^2}{h}\epsilon_{abc}\frac{K^c}{2\pi}, \quad \text{where} \quad \frac{K_c}{2\pi} = \frac{h}{e}\frac{\partial n}{\partial B}\bigg|_{\mu,T=0}. \tag{16}$$

This is Strèda formula. The density n is computed as a the sum over the occupied bands below the Fermi level

$$n = \sum_\alpha \int d^2k \, \det(J(\vec{k}))n_\alpha(\vec{k}). \tag{17}$$

The matrix $J_{ab}(\vec{k})$ is defined in equation 10, and $\det[J(\vec{k})] = 1 + \epsilon_{abc}\frac{e}{\hbar}B^cF^b(\vec{k})$. For a pure Dirac cone $\partial n/\partial B$ changes sign as the chemical potential crosses the Dirac point giving an ambipolar character to the system as a function of chemical potential. In the ARPES spectra, we saw that there was also an additional source of holes coming from the bulk heavy-hole band $\Gamma_{8,HH}$. This is illustrated in Figure 10, where the longitudinal resistivity and the tangent of Hall angle normalized by the

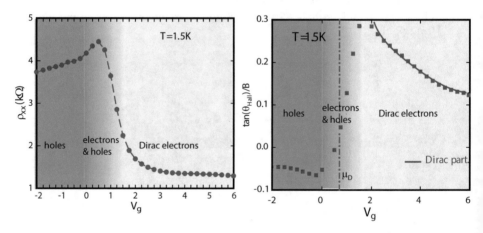

FIGURE 10. Left: longitudinal resistance as a function of gate voltage. Right: Hall angle as a function of gate voltage. Above $2V$, a $1/(V_g - V_{gD})$ divergence is observed as expected for pure Dirac carriers.

field are plotted as a function of gate voltage (chemical potential). The tangent of the Hall angle is the ratio between the Hall and the longitudinal resistivity. For a massive band it is $\tan\theta_H = \omega_c\tau = \frac{e\tau}{m}B$. When normalized by the field it is a constant independent of the chemical potential. For Dirac particles, $\tan\theta_H = 2eD\frac{B}{\mu}$, it is inversely proportional to the chemical potential measured with respect to the Dirac point. We see that for increasing gate voltages, the longitudinal resistivity

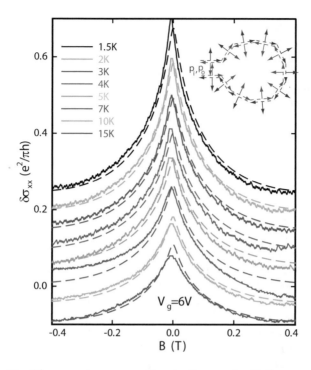

FIGURE 11. The quantum correction to the conductivity are obtained by subtracting the fit to the measured longitudinal conductivity. The difference are plotted as a function of magnetic field for different temperatures. The curves are fitted to the expected digamma dependence as a function of field. The characteristic field is $B_i = 0.04\,\mathrm{mT}$ T=1.5 K and increase with increasing temperature. Beyond B_i, the fitted curves (dotted lines) are dominated by the logarithmic tails expected in 2D.

decreases, while the $(\tan\theta_H)/B$ decrease as $1/\mu \propto 1/(V_g - V_{gD})$, which is the behavior expected for Dirac electrons. As the gate voltage are decreased, the longitudinal resistivity goes through a maximum and the Hall angle switches sign, indicating that the dominant carriers are holes. At negative gate voltages, the longitudinal resistivity only decreases weakly. This suggest a poor coupling to the bulk heavy-holes which have poor mobilities. This ambipolar character of the electronic transport could also be observed in a semimetal, but the $1/\mu$ dependence of the Hall angle at positive gate voltages is a convincing signature of Dirac-like charge carriers in the gapped region.

The last manifestation of the topological nature of the charge carrier is in the quantum interferences present in the magnetoconductance traces.

A resistance can be expressed as a probability of return to the origin for charge carriers [21]. In two dimensions, this probability depends on closed-loops paths. There are two "time-reversed" directions along which charge particles can travel along each closed loop. For loop sizes smaller than the phase coherence length, the propagation amplitudes add coherently. Depending on their relative phase, this can increase or decrease the probability of return to the origin. For topological insulator surface states, the spin stays perpendicular to the momentum after a scattering event (spin-orbit interactions, see Figure 11-inset). After the sequence of scattering on a closed loop, the spin has undergone a 2π rotation, which affects the accumulated phase (sign change): For a given loop, the return probability is proportional to $|u_i(\vec{p},\uparrow) + \Theta u_i(-\vec{p},\downarrow)|^2$, where Θ represents the time-reverse operation: since $\Theta u_i(-\vec{p},\downarrow) = -e^{2\pi\Phi/\Phi_0} u_i(\vec{p},\uparrow)$, the return probability proportional to $\sin^2 2\pi\frac{\Phi}{\Phi_0}$. The sign change comes from the spin rotation and the phase factor $e^{2\pi\Phi/\Phi_0}$ is the accumulated Aharonov–Bohm phase along the loop (Φ is the flux through the loop and $\Phi_0 = h/e$ the flux quantum). We see that the return probability proportional is minimal for zero magnetic field: hence the quantum correction to the conductivity of the Dirac surface states are expected to be *negative*, which is the opposite sign compared to ordinary conductors. Such "anti-localization" quantum corrections to the conductivity have been observed in graphene [22] and other Dirac matter compounds [23] and reveal the presence of a Dirac point [24]. These quantum corrections to the conductivity are obtained by subtracting a fit to the measured longitudinal conductivity and plotted in Figure 11. Their magnetic field dependence can be fitted to the known dependence [24, 27]. At 1.5 K, the magnitude of the weak-localization correction are ≈ 0.5 times smaller than the expected magnitude $e^2/(2\pi h)$ for a perfect Dirac cone. Only anti-localization corrections are observed, ruling out additional contributions (magnetic impurities). As a function of gate voltage V_g the magnitude of quantum corrections drop by a factor of 2.5 and B_i increase by the same factor as ℓ_{so} is reduced by the population of bulk heavy holes which diffuse the surface states.

5. Conclusions

Topological insulators are a distinct class of materials which differs from ordinary band insulators, by the presence of conducting surface states, which wrap around all the outer surface of the material. This 2D-conducting surface sheet differs from other known 2D electron gases found in semiconducting heterostructures, quantum wells or at the surface of metals, in the energy-momentum dispersion $\epsilon(\vec{k}) = \hbar c|k|$ which is relativistic, i.e., linear in momentum $\vec{k}$. In addition, the spin of charge carriers are perpendicular to their momentum, giving a helical spin texture on this Dirac cone, a notable difference with graphene, where 2 Dirac cones are presents which have both possible spin orientations for every momentum.

These surface states have been first studied using angle resolved photoemission (ARPES) and scanning tunnel microscopy (STM), which are powerful surface sensitive probes. These techniques have confirmed the presence of relativistic surface bands, with a helical spin texture. Transport experiments have been successful in material with low bulk conductivity: because of material imperfections, bulk bands often induce a finite conductivity which competes with the surface state contributions. Nevertheless, as materials improves, the typical signature of Dirac carriers are observed in a growing number of experiments.

The helical spin texture of the Dirac cone of topological surface states offers new and interesting scientific perspectives: by selecting the momentum of the carriers, the spin direction is automatically selected. This can be achieved most easily in the spin-Hall regime [25] of 2D topological insulators, where the carriers propagate forward or backward with opposite spin directions at the edge of the 2D structure. Spin filtering [26] and detection using gated H-structures have already been experimentally demonstrated. This spin selection technique with a simple gate offers the possibility to achieve spintronics without the use of magnetic materials.

A final twist which is generating an intense experimental and theoretical effort is the search for Majorana fermions in topological superconductors [2], which can be realized by placing a topological insulator in proximity with an ordinary superconductor. This exciting research goes beyond the topological insulator topic, and gives a new trend in condensed matter science: the combination of materials gives new systems with carefully engineered properties.

References

[1] Hasan, M.Z., and Kane, C.: Rev. Mod. Phys. **82**, 3045 (2010).

[2] Qi, X.-L., and Zhang, S.-C.: Rev. Mod. Phys. **83**, 1057 (2011).

[3] Fu, L., and Kane, C.: Phys. Rev. B **76**, 045302 (2007).

[4] Shockley, W.: Phys. Rev. **56**, 317 (1939).

[5] Tamm, I.: Zeitschrift für Physik **76**, 849 (1932).

[6] Bernevig, B.A., and Zhang, S.-C.: Phys. Rev. Lett. **96**, 106802 (2006).

[7] Winkler, R.: *Spin Orbit Coupling Effects in Two-Dimensional Electron and Hole Systems.* Springer Tracts in Modern Physics, vol. 191, Springer 2003.

[8] Bir, G.L., and Pikus, G.E.: Symmetry and Strain-induced Effects in Semiconductors. Translated from Russian by P. Shelnitz. Translation edited by D. Louvish. New York, Wiley 1974.

[9] Damascelli, A.: Physica Scripta **T109**, 61 (2004).

[10] Ballet, P., Thomas, C., Baudry, X., Bouvier, C., Crauste, O., Meunier, T., Badano, G., Veillerot, M., Barnes, J.P., Jouneau, P.H., et al.: Journal of Electronic Materials pp. 1–8 (2014).

[11] Crauste, O., Ohtsubo, Y., Ballet, P., Delplace, P., Carpentier, D., Bouvier, C., Meunier, T., Taleb-Ibrahimi, A., Lévy, L.P.: Phys. Rev. B **90** (2014), to appear.

[12] Novik, E., Pfeuffer-Jeschke, A., Jungwirth, T., Latussek, V., Becker, C., Landwehr, G., Buhmann, H., and Molenkamp, L.: Phys. Rev. B **72**, 035321 (2005).

[13] Hsieh, D., Xia, Y., Wray, L., Qian, D., Pal, A., Dil, J.H., Osterwalder, J., Meier, F., Bihlmayer, G., Kane, C.L., et al.: Science **323**, 919 (2009).

[14] Hsieh, D., Wray, L., Qian, D., Xia, Y., Dil, J.H., Meier, F., Patthey, L., Oster-walder, J., Bihlmayer, G., Hor, Y.S., et al.: New J. Phys. **12**, 125001 (2010).

[15] Kimura, A., Krasovskii, E.E., Nishimura, R., Miyamoto, K., Kadono, T., Kanomaru, K., Chulkov, E.V., Bihlmayer, G., Shimada, K., Namatame, H., et al.: Phys. Rev. Lett. **105**, 076804 (2010).

[16] Wang, Y.H., Hsieh, D., Pilon, D., Fu, L., Gardner, D.R., Lee, Y.S., and Gedik, N.: Phys. Rev. Lett. **107**, 207602 (2011).

[17] Park, S., Han, J., Kim, C., Koh, Y., Kim, C., Lee, H., Choi, H., Han, J., Lee, K., Hur, N., et al.: Phys. Rev. Lett. **108**, 046805 (2012).

[18] Scholz, M.R., Snchez-Barriga, J., Marchenko, D., Varykhalov, A., Volykhov, A., Yashina, L.V., and Rader, O.: Phys. Rev. Lett. **110**, 216801 (2013).

[19] Haldane, F.: Phys. Rev. Lett. **93**, 206602 (2004).

[20] Fuchs, J.N., Piechon, F., Goerbig, M.O., and Montambaux, G.: Eur. Phys. J. B **77**, 351 (2010).

[21] Montambaux, G.: In Les Houches Summer School, Session LXXXI, edited by H. Bouchiat, Y. Gesen, S. Guéron, G. Montambaux, and J. Dalibard, Elsevier, Amsterdam, Vol. 81, 2005.

[22] Wu, X.S., Li, X.B., Song, Z.M., Berger, C., and de Heer, W.A.: Phys. Rev. Lett. **98**, 136801 (2007).

[23] Chen, J., Qin, H., Yang, F., Liu, J., Guan, T., Qu, F., Zhang, G., Shi, J., Xie, X., Yang, C., et al.: Phys. Rev. Lett. **105**, 176602 (2010).

[24] Kechedzhi, K., McCann, E., Fal'ko, V.I., Suzuura, H., Ando, T., and Altshuler, B.L.: Eur. Phys. J. Special Topics **148**, 39 (2007).

[25] Konig, M., Wiedmann, S., Brune, C., Roth, A., Buhmann, H., Molenkamp, L.W., Qi, X.L., and Zhang, S.C.: Science **318**, 766 (2007).

[26] Brne, C., Roth, A., Buhmann, H., Hankiewicz, E.M., Molenkamp, L.W., Maciejko, J., Qi, X.-L., and Zhang, S.-C.: Nat. Phys. **8**, 485 (2012).

[27] Tkachov, G., and Hankiewicz, E.M.: Weak antilocalization in HgTe quantum wells and topological surface states: massive versus massless Dirac fermions. Phys. Rev. B **84**, 035444 (2011).

Laurent Lévy
Institut Néel
Université Grenoble Alpes and CNRS
BP166
F-38042 Grenoble cedex 9, France
e-mail: `laurent.levy@neel.cnrs.fr`

Dirac Matter, 95–129
© 2016 Springer Basel AG

Topology of Bands in Solids:
From Insulators to Dirac Matter

David Carpentier

Abstract. Bloch theory describes the electronic states in crystals whose energies are distributed as bands over the Brillouin zone. The electronic states corresponding to a (few) isolated energy band(s) thus constitute a vector bundle. The topological properties of these vector bundles provide new characteristics of the corresponding electronic phases. We review some of these properties in the case of (topological) insulators and semi-metals.

1. Introduction

Topology is a branch of mathematics aiming at identifying properties of various objects invariant under continuous deformations. Examples range from the case originally considered by Euler of paths in the city of Königsberg crossing all of its seven bridges, to the two classes of two-dimensional surfaces characterized by an Euler topological index, and the problem of the hairy ball or Möbius strip related to the more abstract vector bundles considered in this paper. Topological considerations occurred in different domains of physics, from the seminal work of P. Dirac on the quantization of magnetic monopoles [1] to the classification of defects such as vortices, dislocations or skyrmions in ordered media [2]. In the last ten years topology has arisen as a useful tool in the old domain of the quantum band theory of crystals. In this theory, single electron states are determined from the geometry of the lattice and the nature of atomic orbitals constituting the solid. These Bloch states are labelled by their quasi-momentum which belongs to the first Brillouin zone, a d-dimensional torus. Mathematically, the object defined out of the ensemble of electronic states labelled by the momentum k is a vector bundle over the Brillouin zone. Such objects are known to display possible non-trivial topology, such as the above-mentioned twisted Möbius strip or the hairy ball. In the context of band theory, this new characterization of an ensemble of eigenstates over the Brillouin torus has led to the notion of topological insulator. The full ensemble of states over the Brillouin torus is always trivial and lacks any topological property. However, it can be split into two well-separated sub-ensemble of states by an energy gap, both of which possessing a non-trivial topology. This is the case of the valence and conduction bands of a topological insulator.

This notion of topological property of bands was first identified in the context of the two-dimensional Integer Quantum Hall Effect soon after its discovery, in the pioneering work of D. Thouless et al. [3]. The initial framework of the band theory of crystals was found to be inadequate and restrictive (in particular translations on the crystal do not commute due to the magnetic field and require the definition of a magnetic unit cell) and soon the initial topological characterization was generalized to a more general form [4, 5]. In this context it later evolved as a property of various interacting electronic and magnetic phases leading to the notion of topological order [6]. The initial requirement of a magnetic field was later lifted by D. Haldane. By considering a model of graphene with a time reversal symmetry breaking potential, but a vanishing net flux through each unit cell, he preserved the translational symmetry of the original lattice [7]. The original Chern topological index of D. Thouless et al. [3] was then understood as a topological property of bands in two-dimensional insulators without time reversal symmetry. These phases are now called Chern insulators. From the beginning, this topological property was discussed in relation with geometrical phases acquired during an adiabatic evolution studied by M. Berry [8]. Both were discussed as the parallel transport of eigenstates following similar "Berry connections" [9]. The topological Chern index describes the impossibility to perform such a parallel transport of Bloch states globally over the Brillouin zone.

The seminal work of C. Kane and G. Mele completely renewed the interest on the topological characterization of bands in crystals and opened new fascinating perspectives of experimental relevance [10]. The authors considered a model of graphene in the presence of a strong spin-orbit interaction, which preserves time-reversal symmetry and discovered a insulating phase characterized by a new topological property, now called the Quantum Spin Hall Effect. While their model takes its roots in the work by D. Haldane, the new topological index is not related in general to the previous Chern index: it characterizes different topological properties. Indeed, the topological twist of a band probed by this Kane–Mele index is a consequence of the constraints imposed on the electronic states by the time-reversal symmetry. It is a symmetry-related topological property. The interest of this new band property grew considerably when it was realized that it existed also in bulk three-dimensional materials. A simple recipe of band inversion due to a strong spin orbit led to the proposal and discovery of this topological property in a large class of materials. How is this topological property detected ? The hallmark of topological bands in an insulator is the existence of metallic states at its surface which can be probed by various surface techniques. Moreover, the surface states are typically described as Dirac particles, whether in one of two dimensions. Indeed, their dispersion relation corresponds to the linear crossing of two bands, which is described phenomenologically at low energy similarly to graphene by a relativistic Dirac equation. The existence of these Dirac surface states implies transport properties different from those of conventional metals, stimulating a great number of theoretical and experimental work.

The present review does not aim to cover all aspects of topological insulating phases, nor to discuss exhaustively any of them. There already exist several detailed reviews [11, 12] and textbooks [13, 14]. Instead we will give a synthetic overview of the salient features of the topological properties of electronic bands. We will start by reviewing the basics of Bloch theory and the Berry phase as a notion of parallel transport of electronic states. We will then discuss the specificities of this Berry formalism within Bloch theory. The definition of the Chern and Kane–Mele topological indices will be derived on simple historical graphene models. We will conclude by a discussion of Dirac fermions both as surface states but also critical models from topological insulators and describe their topological properties.

2. Bloch theory

Crystalline solids are grossly classified into insulators and metals depending on their electronic transport properties. Within the band theory of crystals, this behavior depends on the existence of a gap between energy bands corresponding to occupied electronic states, and empty states for energies above the gap. This notion of energy bands originates from Bloch theory [15]. Indeed, the description of the quantum states of electrons in solids starts by the identification of conserved quantities and associated quantum numbers, in a standard fashion in quantum mechanics. In crystals, the symmetry of the lattice implies that a discrete ensemble of translations commutes with the Hamiltonian, allowing to define a conserved pseudo-momentum k. More precisely, we shall consider electronic wave functions restricted to a crystal $\mathcal{C}$, the locations of atoms, which constitutes a discrete ensemble of points in the Euclidean space of dimension d. By definition, this crystal is invariant by an ensemble of discrete translations T_γ of vectors γ belonging to the so-called Bravais lattice Γ. In general, the initial crystal may not be itself a Bravais lattice: only a subset of translations on the lattice commute with the Hamiltonian. In that case, we can identify $N > 1$ sub-lattices defined as ensembles of points of the crystal $\mathcal{C}$ related by translations of the Bravais lattice. All points of the crystal can then recovered by translations of points of a fundamental domain $\mathcal{F}$, also called a unit cell of the crystal. Necessarily, the choice of a fundamental domain $\mathcal{F}$ amounts to choose one point in each sub-lattice. Of course any translation $\mathcal{F} + \gamma$ of a fundamental domain is also a possible choice for a fundamental domain. However, there exists (for $N > 1$) different choices of $\mathcal{F}$ that are not related by translation: we will come back to this point later in our discussion. A canonical example in two dimensions of such a non-Bravais lattice is the honeycomb lattice of p_z orbitals of Carbon atoms in graphene which we will use throughout this article. It possesses $N = 2$ sub-lattices, represented in Figure 1. Three different choices for fundamental domains for graphene are illustrated in Figure 1.

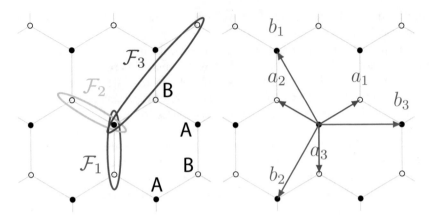

FIGURE 1. Representation of the honeycomb lattice of Carbon atoms in graphene. This lattice is not a Bravais lattice, and possesses 2 sub lattices A and B, represented by black and open circles. Different inequivalent choices of fundamental domains $\mathcal{F}_i$ are shown. Vectors of the triangular Bravais lattice are defined in red on the right figure.

Having determined the Bravais lattice Γ, we can now simultaneously diagonalize the Hamiltonian H defined on the crystal and the set of translation operators T_γ for $\gamma \in \Gamma$. Eigenfunctions of the unitary operators T_γ are Bloch functions, defined by their pseudo-periodicity

$$T_\gamma \psi(x) = \psi(x - \gamma) = e^{ik\cdot\gamma}\psi(x) \text{ for any point } x \in \mathcal{C}, \tag{1}$$

for some vector $k \in \mathbb{R}^d$. This property defines Bloch functions on $\mathcal{C}$ with quasi-momentum k. The ensemble of Bloch functions with a fixed quasi-momentum form a finite-dimensional space $\mathcal{H}_k$ with the scalar product $\langle\psi_k|\chi_k\rangle_k = \sum_{x\in\mathcal{C}/\Gamma} \overline{\psi}_k(x) \chi_k(x)$. At this stage, it is useful to introduce the reciprocal lattice $\Gamma^\star$ of the Bravais lattice Γ, and composed of vectors G such that $G \cdot \gamma \in 2\pi\mathbb{Z}$ for any $\gamma \in \Gamma$. As $e^{ik\cdot\gamma} = e^{i(k+G)\cdot\gamma}$ for $G \in \Gamma^\star$, the spaces $\mathcal{H}_k$ and $\mathcal{H}_{k+G}$ can be identified: the quasi-momentum k takes values only in the d-dimensional Brillouin torus $\mathrm{BZ} = \mathbb{R}^d/\Gamma^\star$ obtained by identifying k and $k + G$ for $G \in \Gamma^\star$.

We consider electronic wave functions belonging to the Hilbert space $\mathbf{H} = \ell^2(\mathcal{C}, \mathbf{V})$ of functions defined on the crystal $\mathcal{C}$, taking values in a finite-dimensional complex vector space V which accounts for the various orbitals per atom kept in the description, and square-summable with the scalar product

$$\langle\psi|\chi\rangle = \sum_{x\in\mathcal{C}}\langle\psi(x)|\chi(x)\rangle_V.$$

The identification of the initial Hilbert space $\mathbf{H}$ with the ensemble of Bloch functions $\mathcal{H}_k$ is provided by the standard Fourier transform. From any such function

$\psi \in \mathbf{H}$, the Fourier transform

$$\widehat{\psi}_k(x) = \sum_{\gamma \in \Gamma} e^{-ik \cdot \gamma} \psi(x - \gamma), \qquad (2)$$

defines a function $\widehat{\psi}_k$ which is pseudo-periodic in the sense defined above:

$$\widehat{T_\gamma \psi}_k(x) = \widehat{\psi}_k(x - \gamma) = e^{ik \cdot \gamma} \widehat{\psi}_k(x). \qquad (3)$$

Note that we also have $\widehat{\psi}_k = \widehat{\psi}_{k+G}$ for $G \in \Gamma^\star$. Hence the function $\widehat{\psi}_k$ belongs to the space $\mathcal{H}_k$ of Bloch functions with pseudo-momentum k. The inverse Fourier transform is naturally given by the integration over the Brillouin torus:

$$\psi(x) = |\mathrm{BZ}|^{-1} \int_{\mathrm{BZ}} \widehat{\psi}_k(x) \, dk, \qquad (4)$$

where $|\mathrm{BZ}|$ stands for the volume of the Brillouin zone. This Fourier transform (2) is an isomorphism between the Hilbert space $\mathbf{H}$ and the ensemble $\{\mathcal{H}_k\}_{k \in \mathrm{BZ}}$ of Bloch functions with momentum k in the Brillouin torus, which allows to "identify" both spaces. The correspondence of norms is provided by the Plancherel formula. The commutation of the Hamiltonian with the translations T_γ for $\gamma \in \Gamma$ implies that the eigenstates ψ_k^α are Bloch functions with k a conserved quantity (quantum number). The evolution of an eigenenergy ϵ_k^α with the quasi-momentum k defines an *energy band*. The number M of energy bands is thus given by $\dim(\mathcal{H}_k) = N \times \dim(V)$: the number of orbitals per lattice site times the number N of sub lattices. The nature of the electronic phase is then determined from both the band structure and the filling factor: how many electrons are initially present per lattice site. Three different situations occurs:

Insulator: An energy gap occurs between occupied states below the chemical potential and empty states above the chemical potentiel. This corresponds to an insulator, with a minimum energy necessary to excite electrons in the crystal

Metal: The chemical potential crosses some energy bands over the Brillouin torus: no minimum exists for the excitation energy

Semi-Metal or Zero Gap Insulators: Two or more energy bands crosses exactly at (or close to) the chemical potential. This is the situation of graphene. Close to this crossing, electronic excitation possess a linear dispersion relation, which mimics that of ultra-relativistic particles.

In the following, we will be interested mostly by the behavior of the eigenstates ψ_k^α for the occupied bands of a crystal, or the bands below the crossing point of semi-metals.

3. Geometrical phase and parallel transport

We now turn ourselves to a discussion of transport of Bloch functions as their momentum k evolves in the Brillouin zone, starting from the initial notion of geometrical phase in quantum mechanics.

3.1. Aharonov–Bohm effect

Consider a small metallic ring threaded by a magnetic field localized inside its hole (see Figure 2). We assume that the propagation of electrons can be considered phase coherent through the ring. Even though the electrons propagating inside

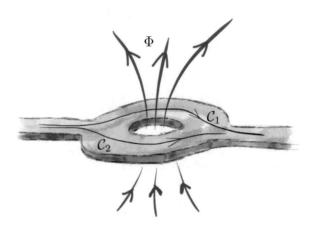

FIGURE 2. Sketch of the geometry of small metallic ring threaded by a magnetic flux, allowing to probe the Aharonov–Bohm effect.

the conductor never encounter the magnetic field, they feel its presence through a dephasing effect: an electron propagating along a trajectory $\mathcal{C}$ acquires an extra phase induced by the electromagnetic potential $A(x)$:

$$\theta_{\mathcal{C}} = -\frac{e}{\hbar} \int_{\mathcal{C}} A(x).dx. \tag{5}$$

This is the so-called Aharonov–Bohm effect [16]: Two trajectories $\mathcal{C}_1$ and $\mathcal{C}_2$ from x_1 to x_2 on opposite sides of the ring will correspond to a respective dephasing

$$\theta_1 - \theta_2 = \frac{e}{\hbar} \oint_{\mathcal{C}_1 \cup \mathcal{C}_2} A(x).dl = 2\pi \frac{\phi}{\phi_0}, \tag{6}$$

where $\phi_0 = h/e$ is the flux quantum for electrons and ϕ is the magnetic flux enclosed between $\mathcal{C}_1$ and $\mathcal{C}_2$. The modulation with the flux ϕ of this dephasing can be experimentally tested by monitoring the electronic current through such a small gold annulus as a function of the magnetic field threading the sample: oscillations of periodicity ϕ_0 are the hallmark of this Aharonov–Bohm effect [17]. Note that although the physical quantity (the current) depends only on the gauge invariant magnetic flux, the evolution along a given path $\mathcal{C}$ depends on the choice of gauge to express the vector potential $A(x)$. A consistent choice is required to be able to describe the evolution of electronic states along any path.

3.2. Berry phase

Let us now discuss the notion of Berry phase. For the sake of pedagogy, we will consider the adiabatic evolution of eigenstates of a Hamiltonian following the initial work of M. Berry [8] (see also [18]), although this notion extends to more general cyclic evolution of states in Hilbert space [19, 20]. We consider a Hamiltonian $H(\lambda)$ parametrized by an ensemble of external parameters $\lambda = (\lambda_1, \lambda_2, \dots)$. These external parameters evolve in time. This continuous evolution can be represented by a curve $\lambda(t), t_i \leq t \leq t_f$ in the parameter space. We study the time evolution of a given non-degenerate eigenstate $|\psi(\lambda)\rangle$ of $H(\lambda)$ with energy $E(\lambda)$. It is further assumed that during the evolution, the energy $E(\lambda)$ is separated from all other energies, and furthermore that the system remains in this particular eigenstate. With these assumptions, the evolution of the initial state takes place in the one-dimensional vector space of instantaneous eigenvectors of $H(\lambda)$ with energy $E(\lambda)$. For normalized states, the only remaining degree of freedom characterizing this evolution lies in the phase of the eigenstate.

To describe the evolution of this phase along the curve $\lambda(t)$ in parameter space, let us consider a "continuous" basis of eigenstates $\{|\phi_n(\lambda)\rangle\}_n$ of the Hilbert space. This basis is assumed continuous in the sense that the $\phi_n(\lambda; x)$ are continuous wave functions as λ varies for any position x. By definition, the eigenstate $|\psi(\lambda(t))\rangle$ can be decomposed on this basis of eigenstates leading to

$$|\psi(\lambda(t))\rangle = e^{-i\theta(t)}|\phi_{n_0}(\lambda(t))\rangle, \tag{7}$$

which satisfies the instantaneous Schrödinger equation

$$H(\lambda(t))\,\psi(\lambda(t))\rangle = i\hbar\,\partial_t|\psi(\lambda(t))\rangle \tag{8}$$

$$\Rightarrow \hbar\,\partial_t\theta(t) = E_{n_0}(\lambda(t)) - i\,\langle\phi_{n_0}(\lambda(t))|\partial_t\phi_{n_0}(\lambda(t))\rangle. \tag{9}$$

The evolution of the phase $\theta(t)$ of the eigenstate is the sum of two contributions: the usual dynamical phase θ_{dyn} originating from the evolution of the energy of the state and a new contribution, denoted the geometrical Berry phase θ_{n_0}:

$$\theta(t_f) - \theta(t_i) = \theta_{\text{dyn}} + \theta_{n_0} \tag{10}$$

$$\theta_{\text{dyn}} = \frac{1}{\hbar}\int_{t_i}^{t_f} E_{n_0}(\lambda(t))dt \quad ; \quad \theta_{n_0} = -i\int_{t_i}^{t_f} \langle\phi_{n_0}(\lambda(t))|\partial_t\phi_{n_0}(\lambda(t))\rangle\,dt.$$

This new contribution can be rewritten as a purely geometrical expression over the parameter space:

$$\theta_{n_0} = -i\int_{\mathcal{C},\lambda[t_i]\to\lambda[t_f]} \langle\phi_{n_0}(\lambda)|d_\lambda\phi_{n_0}(\lambda)\rangle, \tag{11}$$

where we used differential form notations. Hence this Berry phase does not depends of the rate of variation of the parameters $\lambda(t)$ provided the condition of evolution within a single energy subspace is fulfilled: its contribution originates solely from the path $\mathcal{C}$ along which the systems evolves in parameter space.

By analogy with the Aharonov–Bohm electromagnetic contribution (5), we introduce a quantity playing the role of an electromagnetic potential, and characterizing the source of Berry phase evolution at each point λ of parameter space, defined as the 1-form

$$A_{n_0}(\lambda) = \frac{1}{i}\langle\phi_{n_0}(\lambda)|d_\lambda\phi_{n_0}(\lambda)\rangle = \frac{1}{i}\sum_j\langle\phi_{n_0}(\lambda)|\partial_{\lambda_j}\phi_{n_0}(\lambda)\rangle d\lambda_j. \qquad (12)$$

Note that the normalization of the states $|\phi_n(\lambda)\rangle$ ensures that A is a purely real form, as it should be to enforce that θ_{n_0} is a real phase. Moreover, the Berry phase picked up by an eigenstate in an evolution along a closed loop in parameter space can be deduced from the Berry curvature $F(\lambda)$ analogous to the magnetic field, defined by the differential 2-form:

$$\theta_{n_0} = \oint_{\mathcal{C},\lambda[t_f]=\lambda[t_i]} \langle\phi_{n_0}(\lambda)|d_\lambda\phi_{n_0}(\lambda)\rangle \qquad (13)$$

$$= \int_{\mathcal{S},\partial\mathcal{S}=\mathcal{C}} dA(\lambda) = \int_{\mathcal{S},\partial\mathcal{S}=\mathcal{C}} F \quad \text{with } F = dA, \qquad (14)$$

where $\mathcal{S}$ is a surface in parameter space whose boundary corresponds to $\mathcal{C}$. A priori, the expressions (12), (14) appear to depend on the initial choice of continuous basis $\{|\phi_n(\lambda)\rangle\}_n$. This is not the case. Indeed, let us consider a second choice of continuous basis of eigenstates $\{|\phi'_n(\lambda)\rangle\}_n$. The assumption that the state $|\phi_{n_0}(\lambda)\rangle$ is non degenerate implies that $|\phi'_{n_0}(\lambda)\rangle = e^{if_{n_0}(\lambda)}|\phi_{n_0}(\lambda)\rangle$ where $f_{n_0}(\lambda)$ is continuous. Upon this change of basis, the Berry connection $A_{n_0}(\lambda)$ is modified according to

$$A_{n_0}(\lambda) \to A'_{n_0}(\lambda) = A_{n_0}(\lambda) + df_{n_0}(\lambda). \qquad (15)$$

Hence the 1-form $A_{n_0}((\lambda)$ depends on the exact reference basis used to transport eigenstates as λ evolves, however, the associated Berry curvature F_{n_0} does not, and is an intrinsic property of the space of eigenstates $\phi_{n_0}(\lambda)$.

By inserting a completeness relation into the definition (12), we obtain the following alternative expression for the Berry curvature:

$$F_{n_0}(\lambda) = -\sum_{m\neq n_0} \frac{\langle\phi_{n_0}(\lambda)|\partial_{\lambda_j}\phi_m(\lambda)\rangle\,\langle\phi_m(\lambda)|\partial_{\lambda_l}\phi_{n_0}(\lambda)\rangle}{(E_{n_0}(\lambda) - E_m(\lambda))^2}\,d\lambda_j \wedge d\lambda_l. \qquad (16)$$

A direct consequence of this second expression is that the Berry curvature summed over all eigenstates vanishes for any value of the parameters λ:

$$\sum_n F_n(\lambda) = 0. \qquad (17)$$

In the expression (16) the Berry curvature depends on an apparent coupling term between energy bands $E_m(y)$, while the initial expression (14) depends solely on properties of a given band. This second equation expresses in a manifest manner the contraints imposed on the Berry curvature by the projection of the evolution onto the vector space of a single eigenstate. The closer the energies of other states, the stronger this constraint.

3.3. Berry phase and parallel transport on vector bundles

The tools introduced above to describe the evolution of eigenstates of a Hamiltonian as a function of external parameters admit a natural interpretation in terms of parallel transport in a vector bundle [9]. Indeed, in this context the "Berry connection" had already been introduced by D. Thouless $et\ al.$ [3] to characterize the topological properties of the quantum Hall effect on a lattice. We will naturally use this formalism in the following. Mathematically, the object we consider is a vector bundle, which generalizes the notion of tangent spaces over a manifold [21]. Such a vector bundle is constituted of a based space: at each point of this base space is associated a vector space called the fiber. In the present case the base space is constituted of the manifold Λ of external parameters[1] λ, the complex vector fiber is the Hilbert space $\mathbf{H}$ of the problem, independent of λ. The evolution with time t of the parameters corresponds to a curve $\lambda(t), t \in [t_i, t_f]$ on the base space. A vector $|\psi(\lambda)\rangle$ continuously defined as a function of λ is called a section of the vector bundle. To describe the evolution of vector of the fibers along a curve $\lambda(t)$, we need a prescription called a connection or covariant derivative which generalizes the notion of differential. Such a connection defines how vectors are transported parallel to the base space. In the present case, such a prescription is canonically defined: the fiber $\mathbf{H}$ is independent of the point λ considered and the fiber bundle can be written as $\Lambda \times \mathbf{H}$. A natural choice to define parallel transport of vectors amounts to choose a fixed basis $\{|e_\mu\rangle\}_\mu$ of $\mathbf{H}$, independent of the point λ. A vector is then parallel transported if it possesses fixed components in this basis: this amounts to define a covariant derivative or connection ∇ acting on sections and defined by

$$|\psi(\lambda)\rangle = f^\alpha(\lambda)|e_\alpha\rangle \to \nabla|\psi\rangle = (df^\alpha)|e_\alpha\rangle. \tag{18}$$

The covariant derivative ∇ when acting on a vector tangent to the base space measures the "correction" necessary to the section $|\psi(\lambda)\rangle$ to transport it parallel to the frame $\{|e_\mu\rangle\}_\mu$ in the direction of the vector. Hence parallel transported vectors correspond to a choice of section satisfying $\nabla|\psi(\lambda)\rangle = 0$.

The problem considered in the previous section corresponds to an evolution with λ restricted to a subspace of each fiber: the subspace $\mathcal{E}$ associated with an energy eigenvalue $E_{n_0}(\lambda)$ (see Figure 3). It is assumed that this energy remains separated from the rest of the spectrum by a spectral gap as λ is varied. This allows to define unambiguously a projector onto this subspace $P_\mathcal{E}(\lambda) = |\psi_{n_0}(\lambda)\rangle\langle\psi_{n_0}(\lambda)|$ where $|\psi_{n_0}(\lambda)\rangle$ is a normalized eigenstate. This subspace is constituted of a vector space of rank 1 for each λ: if $|\psi_{n_0}(\lambda)\rangle$ is an eigenstate of $H(\lambda)$, so is $a|\psi_{n_0}(\lambda)\rangle$ where $a \in \mathbb{C}$. The transport in this subspace is naturally defined by the projection in the subspace of the canonical connection

$$^\mathcal{E}\nabla = P_\mathcal{E}\nabla. \tag{19}$$

---

[1]Note that the Berry connection can also be introduced on the projective vector bundle of the initial Hilbert space [19, 20].

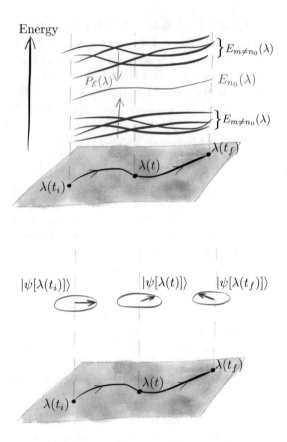

FIGURE 3. Illustration the adiabatic evolution of an eigenstate as a function of external parameters λ as the parallel transport of eigenstates over the manifold of parameters.

This is the so-called Berry connection. A connection is said to be flat if the parallel transport along a loop $\mathcal{C}$ which can be continuously deformed to a point brings any vector back to itself. This is obviously the case for the connection ∇ on the whole vector bundle, but not necessarily for the Berry connection $^{\mathcal{E}}\nabla$. As we will see in the following, only special bundles carry flat connections: this property is associated with a trivialization which is a family of smooth sections $\lambda \mapsto |e^{\alpha}(\lambda)\rangle$, defined over the whole base space and which form an orthonormal basis of every vector fiber. The flatness of a bundle can be locally probed through a curvature 2-form F analogous to the Gaussian curvature for manifolds, and defined as $\nabla^2 |\psi(\lambda)\rangle = F|\psi(\lambda)\rangle$ for any section $|\psi(\lambda)\rangle$. The property of the differential immediately implies that the curvature of the connection (18) vanishes everywhere, as expected for a flat connection. However, this is not the case for the Berry connection.

To express the curvature form of a Berry connection, let us consider again a smooth section $|\psi(\lambda)\rangle = f^\alpha(\lambda)|e_\alpha\rangle$. The Berry connection acting on this section can be expressed as $^\mathcal{E}\nabla|\psi(\lambda)\rangle = \sum_\beta \langle e_\beta|^\mathcal{E}\nabla\psi(\lambda)\rangle \, |e_\beta\rangle$. By using $\nabla f^\alpha(\lambda)|e_\alpha\rangle = df^\alpha(\lambda)|e_\alpha\rangle$ we obtain

$$\langle e_\beta|^\mathcal{E}\nabla|\psi(\lambda)\rangle = \langle e_\beta|P_\mathcal{E}\nabla\left(f^\alpha(\lambda)|e_\alpha\rangle\right) = df^\alpha(\lambda)P^{\beta\alpha}, \tag{20}$$

with $P^{\beta\alpha} = \langle e_\beta|P_\mathcal{E}|e_\alpha\rangle$. The curvature of the Berry connection, or Berry curvature, is a 2-form curvature tensor (analogous to the curvature tensor in the tangent plane of manifolds), whose coefficients

$$F_\mathcal{E}^{\alpha\beta}(\lambda) = \langle e_\alpha|\left(^\mathcal{E}\nabla\right)^2|e_\beta\rangle \tag{21}$$

are expressed as

$$F_\mathcal{E}^{\alpha\beta}(\lambda) = (P_\mathcal{E}\, dP_\mathcal{E}\wedge dP_\mathcal{E})^{\alpha\beta} = P_\mathcal{E}^{\alpha\gamma}dP_\mathcal{E}^{\gamma\delta}\wedge dP_\mathcal{E}^{\delta\beta}, \tag{22}$$

which is often written in a compact form $F_\mathcal{E} = P_\mathcal{E}\, dP_\mathcal{E}\wedge dP_\mathcal{E}$. Its trace $F = \sum_\alpha F^{\alpha\alpha}$ is denoted the scalar curvature (analogous to the Gaussian curvature in the tangent plane of manifolds). To make contact with the initial definition (12), (14) of Berry connection and curvature, let us consider the local frame of the bundle initially introduced and consisting of a basis of eigenstates $\{|\phi_n(\lambda)\rangle\}_n$ of the Hamiltonian $H(\lambda)$. We consider the projector $P_\mathcal{E}$ onto the subspace of eigenvectors of energy $E_{n_0}(\lambda)$. By writing $|\phi_{n_0}(\lambda)\rangle = f_{n_0}^\alpha(\lambda)|e_\alpha\rangle$ we recover the expressions of Berry connection and curvature $A(\lambda) = \langle\phi_{n_0}(\lambda)|^\mathcal{E}\nabla|\phi_{n_0}(\lambda)\rangle = \overline{f_{n_0}^\alpha}(\lambda)df_{n_0}^\alpha(\lambda)$ and $F = dA$. Having established the basic definitions of parallel transport of eigenstates, we can now turn ourselves to its translation in the context of the band theory of solids.

3.4. Bloch vector bundle

In Section 2, we have established that the Bloch theory of electronic states in crystals amounts to identify an ensemble of $\alpha = 1, \ldots, M$ energy bands ϵ_k^α as well as quasi-periodic Bloch functions ψ_k^α. In doing so, we have replaced the initial Hilbert space $\mathbf{H}$ with the collection of complex vector spaces $\mathcal{H}_k$ for k in the Brillouin torus. Mathematically, this collection of vector spaces defines a vector bundle $\mathcal{H}$ over the Brillouin torus BZ, analogous to the vector bundle considered in the context of the Berry phase in the previous section. It is thus tempting to translate the notion of Berry connection and curvature in the context of the band theory of solids. Indeed, such a relation was established by B. Simon soon after the work of M. Berry [9]. Several electronic properties of solids have now been related to these quantities, whose description goes beyond the scope of the present paper (see [22] for a recent review). We will focus on the definitions necessary to the characterization of the topological properties of isolated bands in solids, as well as the specificities of parallel transport in electronic bands in solids, following [23].

The first specificity of the Bloch bundle with respect to the example of Section 3.3 lies in the definition of fibers $\mathcal{H}_k$ which are vector spaces defined specifically at each point k, and thus differs at different points k and k'. In the definition of the

Berry connection, we started from a flat connection over the whole vector bundle, projected onto a vector space of eigenfunctions (see equation (19)). Defining a flat connection is straightforward when we can identify a (local) trivialization of the bundle, *i.e.*, a frame of sections $|e_\alpha^I(k)\rangle$ providing a basis of each fiber $\mathcal{H}_k$. Not all bundles support such a trivialization globally: only those which are called trivial as we will see in the next section. This is indeed the case for the Bloch bundle. However, this property is not obvious from the start: no canonical trivialization exist, as opposed to the situation considered for the Berry connection where the fiber **H** was independent from the point λ in base space. Two such trivializations, also called Bloch conventions (see [24] for a discussion in graphene), are detailed in the appendix, following [23]. The first one amounts to consider the Fourier transform of a basis of functions on a unit cell $\mathcal{F}$, typically localized δ functions. These Fourier transform provide a trivialization of the Bloch bundle: a smooth set of sections which constitute a basis of each fiber $\mathcal{H}_k$. The associated connection ∇^I as well as the associated Berry connections and curvatures generically depends on the initial choice of fundamental domain $\mathcal{F}$. This choice of trivialization is useful when eigenstates and the Bloch Hamiltonian $H(k)$ are required to be periodic on the Brillouin torus, *e.g.*, in the geometrically interpretation of the topological properties of bands (see, *e.g.*, [25]) or when studying symmetry properties [26].

A second choice of trivialization leads to a more canonical connection: it corresponds to the common writing of the Bloch functions $|\varphi(k)\rangle \in \mathcal{H}_k$ in terms of functions $u(k;x)$ periodic on the crystal (with the periodicity of the Bravais lattice Γ):

$$\varphi(k;x) = e^{-ik\cdot(x-x_0)}u(k;x). \qquad (23)$$

A basis of periodic functions on the crystal $\mathcal{C}$, indexed by the sub lattices, is then "pulled back" according to equation (23) as a trivialization of the Bloch bundle. This connection only depends on the arbitrary choice of the origin of space x_0 in (23). If we choose another origin x_0' then $e_\alpha'^{II}(k;x) = e^{ik\cdot(x_0'-x_0)}e_\alpha^{II}(k;x)$ so that the two connection differ by a closed 1-form: $\nabla'^{II} = \nabla^{II} - i\,dk\cdot(x_0'-x_0)$. Note that as this connection ∇^{II} is flat (with a vanishing curvature $(\nabla^{II})^2 = 0$) the associated parallel transport of any state along a contractible loop brings it back to itself. However, this is not so for the loops winding around the Brillouin torus: as can be seen from the relations (62), (63). The connection ∇^{II} is said to have non-trivial holonomies along these loops. This second connection ∇^{II} is particularly relevant for semi-classical analysis as it is related to the position operator by Fourier transform [15]:

$$(x-x_0)_j\psi(x) \leftrightarrow -i\,\nabla^{II_j}\widehat{\psi}(k;x). \qquad (24)$$

Hence physical quantities related to properties of a Berry connection in crystals are indeed related to the properties of this second connection ∇^{II}.

The necessity to distinguish these two natural connections ∇^I and ∇^{II} lies in their Berry connections and curvatures, which generically differ [23]. This is in sharp contrast with the situation considered previously in Section 3.3 where the

Berry curvature was found to be unique. This difference finds its origin in the different nature of the Bloch and Berry bundles. As we will see in the following, topological properties of bands (sub-bundles) do not depend on the initial choice of flat connection. However, when studying geometrical Berry properties of bands and their relevance in observables, one should pay special attention to be consistent in the use of connection on the Bloch bundle. To be more precise on the non unicity of Berry curvature, let us consider two different trivializations $\{|e_\alpha^{(A)}\rangle\}_\alpha$ and $\{|e_\alpha^{(B)}\rangle\}_\alpha$ of the Bloch bundle (either (i) a choice of type I and II defined above, (ii) two trivializations of types I corresponding to different choices of unit cells $\mathcal{F}$, (iii) two trivializations of types II with different choices of origin x_0). Out of these two trivializations we can define two flat connections $\nabla^{(A)}$ and $\nabla^{(B)}$ following (18), and two Berry connections $^\mathcal{E}\nabla^{(A)}$ and $^\mathcal{E}\nabla^{(B)}$ when projected on the same sub-space $\mathcal{E}$ according to (19). Let us now express the corresponding Berry connection tensors (22) in the same trivialization and compare the corresponding scalar curvature (their trace) $F^{(A)}$ and $F^{(B)}$. The change of basis defines a local unitary matrix U by $|e_\alpha^{(A)}\rangle = U_\alpha^\beta(k)|e_\beta^{(B)}\rangle$. The two projection matrices corresponding to $P_\mathcal{E}$ are related through $P^{(B)}(k) = U(k).P^{(A)}(k).U^{-1}(k)$. The two Berry curvatures are found to differ by [23]

$$F^{(B)} = F^{(A)} + \mathrm{tr}\left(dP^{(A)}(k) \wedge U^{-1}dU - P^{(A)}(k)\,U^{-1}dU \wedge U^{-1}dU\right). \qquad (25)$$

As a consequence, generically operators U and P do not commute: we obtain

- for the trivialization I, a Berry curvature F^I which depends on the choice of unit cell $\mathcal{F}$, and differs from the Berry curvature F^II,
- a uniquely defined quasi-canonical Berry curvature F^II, which is independent of the choice of origin x_0. Indeed a change of origin corresponds to scalar matrix U which commutes with any projector.

This dependance of Berry curvatures on the Bloch convention (or trivialization) was noted in [27] in the context of graphene, and is illustrated on Figure 4: the Berry curvature for the valence band of a model of gapped graphene (model of Boron Nitride) is illustrated for two different choices of unit cell in convention I and for convention II. Details of the analysis can be found in [23].

4. Topological properties of a valence band in an insulator

We can now discuss the notion of topological properties of a band in a crystal, or vector bundle. A fiber bundle is said to be topologically trivial if it can be deformed into the product of its base space times a fixed vector space: this corresponds to the situation encountered in the previous section of bundle supporting a set of basis of the fibers continuous on the whole base space, called a frame of sections. Such a bundle, called trivializable, was shown to support a flat connection, *i.e.*, a connection whose curvature vanishes everywhere. Such a continuous basis is not supported by all vector bundles: when this is the case, the bundle is said to be

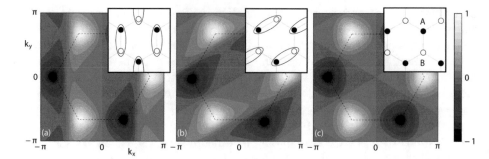

FIGURE 4. The (normalized) Berry curvature $F(k)$ of the valence band in a gapped graphene model, plotted on the Brillouin zone (dashed hexagone) for conventions I (a and b) and II (c). The corresponding choices of unit cells for convention I are shown in the insets. In all three cases, curvature $F(k)$ is concentrated around the Dirac points of ungapped graphene. It depends strongly on the unit cell for convention I and is uniquely defined and respects the symmetries of the crystal for convention II. After [23].

topologically trivial. By definition, a non-trivial vector bundle does not support non-vanishing continuous vector fields (set of sections): they necessarily possess singularities. We will use this property in the following to detect a non-trivial topology of bands in the Bloch bundle. Two examples of non-trivial vector bundles are represented in Figure 5: the Möbius strip, a real bundle on the circle contrasted with the trivial cylinder $\mathcal{S}_1 \times \mathbb{R}$, and a real vector field tangent to the sphere (the hairy ball problem). In both cases, a continuous vector field defined on the bundle necessarily possesses a singularity.

FIGURE 5. Illustration of topology of vector bundles. Left: a trivial (the cylinder) and a non-trivial (the Möbius strip) vector bundle defined as real bundle on the circle $\mathcal{C}_1$. Right: trivial and non-trivial bundles of tangent vectors on the two-dimensional torus and the sphere.

From the discussion in the previous section, we know that the Bloch bundle associated to all energy bands is trivial: it supports a continuous global basis.

A non-trivial topology can only arise in sub-ensembles of bands (sub-bundles). This is indeed the case in an insulator: the gap splits the spectrum into valence and conduction bands. This allows to split each Bloch space $\mathcal{H}_k$ into the vector space of valence and conduction band states whose corresponding vector bundles can acquire topological properties defining a topological insulator. We will see in the following section two different types of a non-trivial topology of a valence band bundle: the first one associated with the Chern index when no symmetry is present, and the second one associated with the Kane–Mele index when time-reversal symmetry is enforced. For simplicity, we will discuss these properties in two dimensions, on the Haldane and Kane–Mele models. Note that while the Chern index only characterizes two-dimensional insulators, the Kane–Mele index can be generalized to three dimensions.

4.1. Chern topological insulators

4.1.1. Two band insulator.
The simplest insulator consists of two bands: one above and one below the band gap. The description of such an insulator is provided by a 2×2 Bloch Hamiltonian $H(k)$ at each point k of the Brillouin zone and parameterized by real functions $h_\mu(k)$ according to

$$H(k) = h^\mu(k)\sigma_\mu = h_0(k)\sigma_0 + \vec{h}(k) \cdot \vec{\sigma} \qquad (26)$$

in the basis of $\sigma_0 = \mathbb{1}$ and Pauli matrices σ_μ. This can be interpreted as the Hamiltonian of a spin $\frac{1}{2}$ coupled to a magnetic field which depends on point k. If we choose a trivialization of type I of the Bloch bundle (basis of Bloch spaces $\mathcal{H}_k$), see Section 3.4, the real functions h_μ are periodic on the Brillouin torus. The two energy bands, defined as eigenvalues of $H(k)$, are

$$\epsilon_\pm(k) = h_0(k) \pm h(k) \qquad (27)$$

with $h(k) = |\vec{h}(k)|$. An insulator situation implies that the two bands never touch each other: therefore $h(k)$ should never vanish on the whole Brillouin torus for the gap to remain finite. We will focus on this situation. The energy shift h_0 of both energies has no effect on topological properties we will discuss: for simplicity we also set $h_0 = 0$. If we use spherical coordinates to parametrize $\vec{h}(k)$, the eigenvectors of the valence band $\epsilon_-(k)$ below the gap are defined, up to a phase, by

$$\vec{h}(h(k), \phi(k), \theta(k)) = h \begin{pmatrix} \sin\theta \, \cos\varphi \\ \sin\theta \, \sin\varphi \\ \cos\theta \end{pmatrix} \quad \longrightarrow \quad |u_-(\vec{h}(k))\rangle = \begin{pmatrix} -\sin\frac{\theta}{2} \\ e^{i\varphi} \cos\frac{\theta}{2} \end{pmatrix}. \qquad (28)$$

From this expression, we realize that the norm $h = |\vec{h}|$ does not affect the eigenvector $|u_-(\vec{h}(k))\rangle$: therefore, the parameter space describing the vectors of the valence band is a 2-sphere S^2. A continuous vector field $|u_-\rangle$ cannot be defined on S^2: it possesses necessarily singularities (vortices). In other words the corresponding vector bundle on the sphere S^2 is not trivial: the valence band vector bundle on the Brillouin torus is then nontrivial as soon as the function $k \mapsto \vec{h}(k)$ covers the whole sphere. This behavior unveils the nontrivial topology of a vector bundle on the sphere, discovered by Dirac and Hopf in 1931 [1]. It is related to the well-known property that the phase of the electronic state around a magnetic monopole cannot be continuously defined: it is necessarily singular at a point, here the North pole $\theta = 0$ where the state (28) still depends on the ill-defined angle ϕ. Indeed the situation where $\vec{h}(k)$ covers the whole sphere as k varies in the Brillouin torus necessarily corresponds to a situation, if $\vec{h}(k)$ was a magnetic field, where the interior of the torus would contain a monopole. This situation exactly describes a non-trivializable vector bundle: if $\vec{h}(k)$ covers the whole sphere, whatever the change of phase we try on the state (28), it is impossible to find a continuous basis of the valence band bundle defined everywhere.

How to detect in general that the ensemble of eigenstates of bands below the gap possess such a topological property ? If $\vec{h}(k)$ was a magnetic field, we would simply integrate its magnetic flux through the Brillouin torus to detect the magnetic monopole. For more general vector bundles, the quantity playing the role of the magnetic flux is the Berry curvature introduced in Section 3.3: it is the curvature associated with any Berry connection. When integrated over the Brillouin zone (the base space of the vector bundle), it defines the so-called first Chern number

$$c_1 = \frac{1}{2\pi} \int_{\text{BZ}} F. \tag{29}$$

This integral of a 2-form is only defined on a two-dimensional surface: this Chern number characterizes bands of insulators in dimension $d = 2$ only. Let us also notice that this Chern number, which is a topological property of the bundle, is independent of the connection chosen. We have seen in Section 3.4 that different Berry connections with different Berry curvatures could be defined on a Bloch sub bundle: they all give rise to the same value of the integral in equation (29) as can be checked from equation (25).

Let us now calculate this Chern number for the valence band of the model (26): the curvature 2-form takes the form [8]

$$F = \frac{1}{2} \frac{\vec{h}}{|h|^3} \cdot \left(\frac{\partial \vec{h}}{\partial k_x} \times \frac{\partial \vec{h}}{\partial k_y} \right) \, \mathrm{d}k_x \wedge \mathrm{d}k_y, \tag{30}$$

corresponding to a Chern number

$$c_1 = \frac{1}{4\pi} \int_{\text{BZ}} \frac{\vec{h}}{|h|^3} \cdot \left(\frac{\partial \vec{h}}{\partial k_x} \times \frac{\partial \vec{h}}{\partial k_y} \right) \, \mathrm{d}k_x \wedge \mathrm{d}k_y. \tag{31}$$

We recognise in the expression (31) the index of the function $\vec{h}$ from the Brillouin torus to the sphere S^2, which exactly counts the "winding" of this map around the sphere, as expected.

4.1.2. The Haldane model. We can apply this characterization of valence bands to the study of a toy-model defined on graphene, first proposed by D. Haldane [7].

The corresponding first quantized Hamiltonian can be written as:

$$H = t \sum_{\langle i,j \rangle} |i\rangle\langle j| + t_2 \sum_{\langle\langle i,j \rangle\rangle} |i\rangle\langle j| + M \left[\sum_{i \in A} |i\rangle\langle i| - \sum_{j \in B} |j\rangle\langle j| \right] \tag{32}$$

where $|i\rangle$ represents an electronic state localized at site i (atomic orbital), $\langle i,j \rangle$ represents nearest neighbors sites and $\langle\langle i,j \rangle\rangle$ second nearest neighbors sites on the honeycomb lattice (see Figure 1), while A and B correspond to the two sublattices of the honeycomb lattice. This Hamiltonian is composed of a first nearest neighbors hopping term with a hopping amplitude t, a second neighbors hopping term with a hopping parameter t_2, with on-site energies $+M$ on sublattice A, and $-M$ for sublattice B, which breaks inversion symmetry. Moreover, each unit cell of the lattice is threaded by magnetic fluxes which average to zero, but lead to Aharanov–Bohm phases which break time-reversal symmetry, and are taken into account through the Peierls substitution $t_2 \to t_2 e^{i\phi}$ while t is unaffected by the flux. The Fourier transform (according to convention I defined previously) of the Hamiltonian (32) leads to a 2×2 Bloch Hamiltonian $H(k) = h^\mu(k)\sigma_\mu$ in the (A,B) sublattices basis

$$h_0 = 2t_2 \cos\phi \sum_{i=1}^{3} \cos(k \cdot b_i) \; ; \qquad h_z = M - 2t_2 \sin\phi \sum_{i=1}^{3} \sin(k \cdot b_i) \; ; \tag{33a}$$

$$h_x = t\left[1 + \cos(k \cdot b_1) + \cos(k \cdot b_2)\right] \; ; \quad h_y = t\left[\sin(k \cdot b_1) - \sin(k \cdot b_2)\right] \; ; \tag{33b}$$

with a convention where $\vec{h}$ is periodic: $\vec{h}(k + G) = \vec{h}(k)$ and the vectors b_i are defined in Figure 1.

To determine the phase diagram, we first consider the energy difference $|h(k)|$ between the two bands which is positive for all points k in the Brillouin zone except for $|M|/t_2 = 3\sqrt{3}\sin\phi$ where it vanishes at one point in the Brillouin zone, and two points only for the case of graphene $M = \phi = 0$. Hence we obtain the phase diagram of Figure 6 which consists of three gapped phases (insulators) separated by transition lines. Along these critical lines, the system is not insulating anymore: it is a semi-metal with low energy Dirac states. The insulating phases are not all equivalent: their valence bands possess different topological Chern number. As a consequence, the transition between these different insulating phases necessarily occurs through a gap closing transition. These Chern numbers can be determined directly using the expression (31), or in a simpler manner by realizing that the winding of the function $\vec{h}(k)$ around the sphere also corresponds to the algebraic number of crossing a ray originating from the origin (as $\vec{h}(k)$ describes a closed surface Σ on the sphere) [28]. By choosing such a line as the Oz axis, we identify the wavevectors k such that $\vec{h}$ crosses the z axis corresponding to $h_x(k) = h_y(k) = 0$: they correspond to the Dirac points $\vec{K}, \vec{K}'$ of graphene. The Chern number is then expressed as

$$c_1 = \frac{1}{2} \sum_{k=K,K'} \text{sign}\left[h(k) \cdot n(k)\right], \tag{34}$$

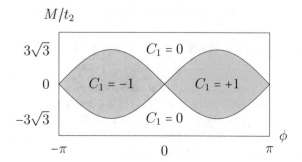

FIGURE 6. Phase diagram of the Haldane model (33) as a function of the Aharonov–Bohm flux ϕ (amplitude of time-reversal breaking term) and alternate potential M/t_2 measuring the inversion breaking amplitude. Insulating phases are characterized by the Chern number c_1.

where $n(k) = \pm\hat{e}_z$ is the normal vector to Σ at k. We obtain

$$c_1 = \frac{1}{2}\left[\ \text{sign}\ \left(\frac{M}{t_2} + 3\sqrt{3}\sin(\phi)\right) - \ \text{sign}\ \left(\frac{M}{t_2} - 3\sqrt{3}\sin(\phi)\right)\right], \qquad (35)$$

which corresponds to the original result of [7]: the corresponding values for the gapped phases are shown on Figure 6.

4.2. Kane–Mele topological index with time-reversal symmetry

4.2.1. Time-reversal symmetry in Bloch bands.
Within quantum mechanics, the time-reversal operation is described by an anti-unitary operator Θ [29], which satisfies $\Theta(\alpha x) = \overline{\alpha}\Theta(x)$ for $\alpha \in \mathbb{C}$ and $\Theta^\dagger = \Theta^{-1}$. When spin degrees of freedom are included, the unitary part of Θ can be written as a π rotation in the spin space $\Theta = e^{-i\pi S_y/\hbar}\ K$, where S_y is the y component of the spin operator, and K is the complex conjugation. Therefore for half-integer spin particles time-reversal operation possesses the crucial property: $\Theta^2 = -\mathbb{I}$. This property has important consequences for the band theory of electrons in crystals. In this context, as time-reversal operation relates momentum k to $-k$, Θ acts as an anti-unitary operator from the Bloch fiber $\mathcal{H}_k$ to $\mathcal{H}_{-k}$. Time-reversal invariance of a band structure implies that the Bloch Hamiltonians at k and $-k$ satisfy:

$$H(-k) = \Theta H(k)\Theta^{-1}. \qquad (36)$$

This equation implies that if $|\psi(k)\rangle$ is any eigenstate of the Bloch Hamiltonian $H(k)$ then $\Theta|\psi(k)\rangle$ is an eigenstate of the Bloch Hamiltonian $H(-k)$ at $-k$, with the same energy. This is the Kramers theorem. Moreover $\Theta^2 = -\mathbb{I}$ implies that these two Kramers partners are orthogonal. In general the two Kramers partners lie in different Bloch fibers $\mathcal{H}_k$ and $\mathcal{H}_{-k}$, except at special points where k and $-k$ are identical on the torus, *i.e.*, differ by a reciprocal lattice vector: $k = -k + G \Leftrightarrow k = G/2$. These points are called time-reversal invariant momenta (TRIM), and

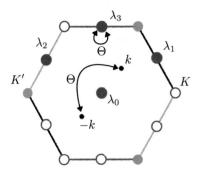

FIGURE 7. Representation of the Brillouin torus of graphene (triangular Bravais lattice) with the K and K' Dirac points, the points k and $-k$ ($+G$) related by time reversal operator Θ and the four time-reversal invariant momenta λ_i.

denoted by λ_i in the following. They are represented on the Figure 7 for the Brillouin torus of graphene. Above these points, the two Kramers partners live in the same fiber $\mathcal{H}_{\lambda_i}$ and the spectrum is necessarily degenerate.

Note that in a time-reversal invariant system of spin $\frac{1}{2}$ particles, the Berry curvature within valence bands is odd: $F_\alpha(k) = -F_\alpha(-k)$. Hence the Chern number of the corresponding bands vanishes: the valence vector bundle is always trivial from the point of view of the first Chern index. It is only when the constraints imposed by time-reversal symmetry are considered that a different kind of non-trivial topology can emerge. The origin of this new topological index can be understood as follows: to enforce the constraints imposed by time reversal symmetry, it appears advisable to define the pairs of spin $\frac{1}{2}$ eigenstates on half of the Brillouin torus. It is always possible to do so in a continuous manner. By application of the time reversal operator Θ, continuous eigenstates are then defined on the second half of the Brillouin torus. However, a smooth reconnection of the eigenstates at the boundary of the two halves Brillouin torus is not guaranteed and not always possible up to a deformation of the initial eigenstates: see Figure 8. The new Kane–Mele topological index aims precisely at measuring this possible obstruction to define continuously Kramers pairs on the whole Brillouin torus. It is intimately related to the constraints imposed by time-reversal symmetry: it will be topological in the sense that the associated properties are robust with respect to all perturbations that preserve this symmetry and the isolation of the band considered. Similar topological indices associated with stronger crystalline symmetries can be defined, whose discussion goes beyond the scope of this paper.

4.2.2. Kane–Mele model. For simplicity, we will restrict ourselves to the discussion of the new $\mathbb{Z}_2$ index in two dimensions, although it can be generalized to $d = 3$ as opposed to the Chern index. We will start with a discussion of the model initially

FIGURE 8. Eigenstates of an isolated band (such as valence bands in an insulator) can be defined continuously on half of the Brillouin zone. When time reversal symmetry is present, the eigenstates on the other half are defined unambiguously by the action of the time reversal operator Θ. The smooth reconnection between the eigenstates of the two halves is not guaranteed: a winding of the relative phase between the two opposite states prevents such a possibility. The Kane–Mele $\mathbb{Z}_2$ topological index probes this possible obstruction to define continuously eigenstates over the Brillouin zone as a consequence of time-reversal symmetry constraints.

proposed by Kane and Mele [10], in a slightly simplified form [30]. This model describes a sheet of graphene in the presence of a particular form of spin-orbit coupling. While the proposed Quantum Spin Hall topological phase has not been detected in graphene due to the small amplitude of spin-orbit in Carbon atoms, it constitutes the simplest model to discuss this new topological index, and shares a lot of common features with the Haldane model of Section 4.1, which justifies our choice to discuss it.

By Kramers theorem, the simplest insulator with spin-dependent time-reversal symmetric band structure possesses two bands of Kramers doublet below the gap, and two bands above the gap: it is described by a 4×4 Bloch Hamiltonian. While such a Hamiltonian is generally parametrized by 5 Γ_a matrices which satisfy a Clifford algebra $\{\Gamma_a, \Gamma_b\} = 2\delta_{a,b}$ and their commutators [30], a minimal model sufficient for our purpose can be written in terms of three such matrices [10, 30]

$$H(k) = d_1(k)\, \Gamma_1 + d_2(k)\, \Gamma_2 + d_3(k)\, \Gamma_3. \tag{37}$$

These gamma matrices are constructed as tensor products of Pauli matrices that represent two two-level systems: in the present case they correspond to the sub lattices A and B of the honeycomb lattice (see Figure 1) and the spins of electrons. In the basis

$$(A, B) \otimes (\uparrow, \downarrow) = (A\uparrow, A\downarrow, B\uparrow, B\downarrow), \tag{38}$$

these matrices read

$$\Gamma_1 = \sigma_x \otimes I \ ; \ \Gamma_2 = \sigma_y \otimes I \ ; \ \Gamma_3 = \sigma_z \otimes s_z, \tag{39}$$

where σ_i and s_i are sub lattice and spin Pauli matrices. The first two terms in (37) are spin independent and describe nearest neighbor hopping of electrons in graphene. $d_1(k), d_2(k)$ accounts for the hopping amplitudes on the honeycomb lattice (see, *e.g.*, equation (33b)). The only term acting on the spins of electrons is the last term, which plays the role of spin-orbit: here this choice of matrix is dictated by simplicity such as to preserve the s_z spin quantum numbers. This not necessary but simplifies the discussion. A general spin-orbit coupling will obviously not satisfy this constraint, but provided it doesn't close the gap it will not modify the following topological properties. With those choices, the time-reversal operator reads:

$$\Theta = i \left(\mathbb{I} \otimes s_y \right) K, \tag{40}$$

while the parity (or inversion) operator exchanges the A and B sub lattices:

$$P = \sigma_x \otimes \mathbb{I}, \tag{41}$$

whose two opposite eigenvalues correspond to the symmetric and antisymmetric combination of A and B components. Let us now enforce the time-reversal symmetry of this model: the commutation rules

$$\Theta\Gamma_1\Theta^{-1} = \Gamma_1 \ ; \ \Theta\Gamma_2\Theta^{-1} = -\Gamma_2 \ ; \ \Theta\Gamma_3\Theta^{-1} = -\Gamma_3 \tag{42}$$

imply that the constraint $\Theta H(k)\Theta^{-1} = H(-k)$ is fulfilled provided $d_1(k)$ is an even and $d_2(k), d_3(k)$ odd functions on the Brillouin torus:

$$d_1(k) = d_1(-k+G) \ ; \ d_2(k) = -d_2(-k+G) \ ; \ d_3(k) = -d_3(-k+G) \ ; \quad \text{for } G \in \Gamma^*. \tag{43}$$

Note that with these constraints, the Bloch Hamiltonian also satisfy $PH(k)P^{-1} = H(-k)$: parity is also a symmetry of the model, with eigenvalues $d_1(\lambda_i)$ at the TRIM points. As a consequence, the $P\Theta$ symmetry of the model implies that the spectrum is degenerate over the whole Brillouin torus and not only at the TRIM λ_i points.

Diagonalisation of the Hamiltonian (37), (39) yields the eigenenergies $\epsilon_\pm(k) = \pm d(k)$ with $d(k) = (d_1^2 + d_2^2 + d_3^2)^{1/2}$. Hence an insulator corresponds to the situation where d_1, d_2 and d_3 cannot simultaneously vanish. We will restrict ourself to this situation. The eigenvectors for the valence band $\epsilon_-(k)$ below the gap can be determined with special care to ensure Kramers degeneracy:

$$|u_{-,\downarrow}\rangle = \frac{1}{\mathcal{N}_1} \begin{pmatrix} 0 \\ -d_3 - d \\ 0 \\ d_1 + id_2 \end{pmatrix} \quad \text{and} \quad |u_{-,\uparrow}\rangle = \frac{1}{\mathcal{N}_2} \begin{pmatrix} d_3 - d \\ 0 \\ d_1 + id_2 \\ 0 \end{pmatrix}, \tag{44}$$

where $\mathcal{N}_j(\vec{d})$ are normalization factors. For these states, singularities appear when $d_1 = d_2 = 0$. Through the polar decomposition $d_1 + id_2 = t\,e^{i\theta}$, we obtain in the limit $t \to 0$:

$$|u_{-,\downarrow}\rangle \to \begin{pmatrix} 0 \\ -1 \\ 0 \\ 0 \end{pmatrix} \quad \text{and} \quad |u_{-,\uparrow}\rangle \to \begin{pmatrix} 0 \\ 0 \\ e^{i\theta} \\ 0 \end{pmatrix} \qquad \text{for } d_3 > 0, \qquad (45a)$$

$$|u_{-,\downarrow}\rangle \to \begin{pmatrix} 0 \\ 0 \\ 0 \\ e^{i\theta} \end{pmatrix} \quad \text{and} \quad |u_{-,\uparrow}\rangle \to \begin{pmatrix} -1 \\ 0 \\ 0 \\ 0 \end{pmatrix} \qquad \text{for } d_3 < 0. \qquad (45b)$$

The phase θ is ill defined when $t \to 0$, so one of these eigenstates is singular in this limit depending on the sign of d_3. In the Brillouin zone, the Kramers theorem imposes that these singularities necessarily occur by pairs $\pm k_0$ of singularity of opposite vorticity for the phase ($d_3(k_0)$ and $d_3(-k_0)$ possessing opposite signs). Note that with our choice of eigenstates (44), these singularities occur exactly at the Dirac points $\pm K$ of graphene, which are defined by $d_1(\pm K) = d_2(\pm K) = 0$. This is a consequence of our choice of eigenstates with a well-defined S_z spin quantum number. The reader can convince himself that when such a pair of singularities exist, its location k_0 can be modified by a general $U(2)$ rotation in the space of eigenvectors[2], i.e., at the expense of preserving the S_z spin quantum number for the eigenstates. (see also [31] for a related point of view).

Thus triviality of the valence band bundle depends on the existence in the Brillouin torus of points where d_1 and d_2 simultaneously vanish. Topologically, this existence depends solely on the sign of d_1 at the four TRIM points λ_i: condition (43) imposes that the odd functions $d_2(k), d_3(k)$ must vanish at the λ_i points, but also along arbitrary lines connection these λ_i points. On the other hand the functions $d_1(k)$ is necessarily non-vanishing at the λ_i points. If the $d_1(\lambda_i)$ all have the same sign, e.g., positive, then a single choice of nonsingular eigenvectors $|u_{-,\uparrow\downarrow}(k)\rangle$ over the whole Brillouin zone is possible: the valence band bundle is trivial. On the other hand if one of the $d_1(\lambda_i)$ have a sign opposite to the other, $d_1(k)$ vanishes on a line encircling this λ_i. Necessarily there exists then two points where both d_1 and d_2 vanish: the valence band bundle is topologically non trivial. If $d_1(\lambda_i)$ takes a negative sign at a pair of TRIM, the singularities can be avoided and the valence band bundle is again trivial. Through this reasoning, we realize that the triviality of the valence band bundle is directly related to the sign of

$$\prod_{\lambda_i} \operatorname{sign} d_1(\lambda_i) = (-1)^\nu. \qquad (46)$$

When this product is positive, the valence band bundle is topologically trivial, while it is not trivial when this product is negative. This defines a new $\mathbb{Z}_2$ topological index ν, called the Kane–Mele index. Morerover this expression can be easily generalized to a three-dimensional model as shown by Fu, Kane, and Mele [32]. It is of practical importance to realize that in this expression, the $d_1(\lambda_i)$ correspond

---

[2]I am indebted to Krzysztof Gawedzki for a clarification on this point.

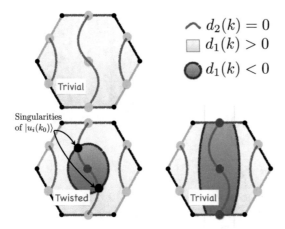

FIGURE 9. Illustration of the condition of existence of singularities for the valence band eigenfunctions $|u_-(k)\rangle$ of the Kane–Mele model. These singularities exist when $d_1(k) = d_2(k) = 0$ which necessarily occurs when $\prod_i d_1(\lambda_i) < 0$: this is the condition which defines the topological non-triviality of these valence bands.

to the parity eigenvalues of the band at the high symmetry points λ_i: a simple recipe to induce a $\mathbb{Z}_2$ topological order is to look for insulators with a *band inversion* around one of the symmetry point, *i.e.*, where the natural order of atomic orbitals composing the bands has been reversed. A band inversion between two orbitals of opposite parity induces a topological twist of the valence bands.

4.2.3. Expression of the Kane–Mele index. The expression (46) for the so-called Kane–Mele index was derived on more rigorous ground for models with parity symmetry, while more general expressions exist when parity is absent. The original derivation proceeds as follows [10]. As the Chern number of the valence band vanishes, there exist a trivialization of the corresponding bundle, *i.e.*, global continuous states $\{|e_\alpha(k)\rangle\}_{\alpha=1,...,2m}$ constituting a basis of the filled band fiber at each point k. Kane and Mele originally considered the matrix element of the time-reversal operator Θ in this basis[3] [10]:

$$m_{\alpha\beta}(k) = \langle e_\alpha(k)|\Theta e_\beta(k)\rangle. \tag{47}$$

This $2m \times 2m$ matrix is not unitary, but it is antisymmetric (because $\Theta^2 = -1$): it possesses a well-defined Pfaffian $\mathrm{Pf}\,(m)$. This quantity tracks the orthogonality between Kramers related eigenspaces of the valence band. This property can be monitored by considering the vortices of this Pfaffian. Time reversal invariance imposes that these vortices necessarily come by pair $k_0, -k_0$ of opposite vorticities.

---

[3]To define the scalar product of vectors between the different the fibers at k and $-k$, we use the trivialization as $\mathrm{BZ} \times \mathbb{C}^{2m}$ of the valence bundle Bloch states and the scalar product in $\mathbb{C}^{2n}$.

FIGURE 10. Phase of the pfaffian Pf (m) on the Brillouin zone for the Kane–Mele model (37), (39) in a trivial (left, $\mu = -3$) and topological insulating phases (right, $\mu = -1$). The parametrizing functions adapted from [30] are $d_1(k_x, k_y) = \mu + \cos k_x + \cos k_y$, $d_2(k_x, k_y) = \sin k_x + \sin k_y$, $d_3(k_x, k_y) = \sin k_x - \sin k_y - \sin(k_x - k_y)$. The middle figure represent the phase color convention.

On the other hand, at any time-reversal invariant points λ_i, the Pfaffian has unit modulus $|\mathrm{Pf}\, m(\lambda_i)| = 1$ and thus cannot support a vortex. A continuous deformation of the vectors constituting the valence band allows to move continuously these vortices in the Brillouin zone, however two situations occur:

- for an even number of vortices, vortices of opposite vorticity can annihilate by pair away from the λ_i points. Topologically this situation is equivalent to the absence of vortex.
- for an odd number of vortices, one pair of opposite vortices remain, which cannot be annihilated through the λ_i points.

These considerations allowed Kane and Mele to define their index ν as the parity of the number of pairs of vortices of Pf $m(k)$ in the Brillouin zone, or more practically the parity of the number of vortices in half of the Brillouin zone:

$$\nu = \frac{1}{2\pi i} \oint_{\partial \frac{1}{2} \mathrm{BZ}} \mathrm{d} \log \left[\mathrm{Pf}\, (m) \right] \quad \mathrm{mod}\ 2. \tag{48}$$

As an illustration, let us apply this expression to the Kane–Mele model (37), (39). In this case, the Pfaffian is: Pf $(m) = d_1(d_1 + i d_2)/((d_1^2 + d_2^2)^{1/2} d)$ which possesses vortices at the points $d_1 = d_2 = 0$ where the valence band eigenstates were previously found to be singular. These vortices as represented in Figure 10. Alternate and more practical expressions of the ν invariant exist: one of them, introduced by Fu and Kane [33], uses the so-called sewing matrix

$$w_{\alpha\beta}(k) = \langle e_\alpha(-k) | \Theta e_\beta(k) \rangle, \tag{49}$$

to define the $\mathbb{Z}_2$ invariant as

$$(-1)^\nu = \prod_{\lambda_i} \frac{\text{Pf } w(\lambda_i)}{\sqrt{\det w(\lambda_i)}}, \tag{50}$$

which simplifies into

$$(-1)^\nu = \prod_{\lambda_i} \text{sign } d_1(\lambda_i) \tag{51}$$

when parity symmetry is present with eigenvalues $d_1(\lambda_i)$ at the TRIM. A different expression formulates explicitly the invariant as an obstruction to defined continuously the phase of Kramers partners [34, 33]. This expression can also be defined from a homotopy point of view allowing for an easy generalization to three dimension[35]. It allows for practical numerical determination of the Kane–Mele topological index [36]. Other expressions of the Kane–Mele $\mathbb{Z}_2$ invariant have also been discussed in relation to a Chern–Simons topological effective field theory [37, 38].

4.3. Insulating materials displaying a topological valence band

The Kane–Mele model discussed above is certainly oversimplified to realistically describe materials possessing a Kane–Mele non-trivial topology. However, its study allowed to identify two crucial ingredients when seeking for topological insulators:

- we should look for materials with a strong spin dependent interaction preserving time-reversal symmetry, and spin-orbit is the obvious candidate.
- a topologic order is associated with the inversion at one of the high symmetry point of the Brillouin zone of two bands of opposite parities. This implies that one of the bands (typically the highest) below the gap should originates around this point from atomic orbitals which usually have a higher energy. Such a *band inversion* is typically associated with a strong spin-orbit coupling.

The band inversion by spin-orbit condition imposes to look for materials with a gap comparable with the weak amplitude of spin-orbit. Hence the materials in which this topological order has been identified and proposed are small gap semiconductors. Several of them such as HgTe/CdTe or InAs/GaSb proposed by the group of S.-C. Zhang [39, 40] to realize the two-dimensional topological phase are commonly used as infrared detectors with energy gaps compatible with the infrared wavelengths. It is important to realize that while the topological index probes a global property of eigenstates over the whole Brillouin zone, the crucial property of formula (46) is that the topological non-triviality can be detected locally at one of the symmetry points! This is a remarkable property of practical importance. Indeed, most studies of topological insulators properties focuses on an effective $k.P$ description of eigenstates close to the minimal gap [41] which is certainly not valid in the whole Brillouin zone, but yet captures all essential ingredients of these new phases. Beyond the quantum wells built out of HgTe/CdTe or InAs/GaSb semiconductors already mentioned, and in which the two-dimensional Quantum

Spin Hall phase has been experimentally detected, canonical three-dimensional topological insulators include the Bi compounds such as Bi_2Se_3, Bi_2Te_3, or bulk HgTe strained by a CdTe substrate.

5. From insulators to semi-metals

5.1. From topology in the bulk to demi-metallic surface states

The essential consequence of a nontrivial topology of valence bands in an insulator is the appearance of *metallic edge states* at its surface, with a linear relativistic dispersion relation. It is these surface states that are probed experimentally when testing the topological nature of a semi-conductor, and which give rise to all physical properties of these materials. The continuous description of the interface between two insulators amounts to continuously deform the bulk eigenstates of one into the other (see [41]). If one of the two insulators is topological this is not possible for the ensemble of valence band eigenstates by definition of the topological nature of the twist. Extrapolation of wave functions from one side to the other requires to close the gap: a smooth connection is then always possible as a consequence of the triviality of the Bloch bundle. Another more practical point of view consists in recalling that the $\mathbb{Z}_2$ topological order can be identified with the band inversion at one of the TRIM point of the Brillouin zone with respect to the "standard ordering" of atomic orbitals constituting bands in a standard insulator. Hence at the interface between two such insulators, the two corresponding bands have to be flipped back from the inverted to the normal ordering. When doing so, they necessarily give rise to energy bands within the gap and located at the interface, corresponding to edge or surface states. Moreover, the crossing of two such bands can be described at low energy by a linear dispersion relation, associated with a relativistic Dirac equation of propagation.

These edge states can be explicitly derived for the two two-dimensional models considered in this article: let us first consider the interface at $y = 0$ between two different insulating phases of the Haldane model (32). At a transition line, the gap m changes sign at one of the two Dirac points of graphene ($h_x(K) = h_y(K) = 0$). Assuming that the Bloch Hamiltonian away from this point K is not modified significantly at the transition, we describe the interface by focusing solely on the low energy degree of freedom and linearize the Bloch Hamiltonian around point $k = K + q$ to obtain a massive Dirac Hamiltonian with mass m:

$$H_1(q) = \hbar v_F \left(q_x \cdot \sigma_x + q_y \cdot \sigma_y \right) + m(y)\, \sigma_z, \tag{52}$$

with $m = h_z(K + q, y)$ such that $m(y < 0) < 0$ and $m(y > 0) > 0$. In real space, this Dirac equation reads (with $\hbar v_F = 1$):

$$H_1 = -\mathrm{i}\, \nabla \cdot \sigma_{2d} + m(y)\sigma_z. \tag{53}$$

Solving this equation, we obtain a state of energy $E(q_x) = E_F + \hbar v_F q_x$ localized at the interface $y = 0$ [42]:

$$\psi_{q_x}(x, y) \propto e^{iq_x x} \exp\left[-\int_0^y m(y')\,dy'\right]\begin{pmatrix}1\\1\end{pmatrix}. \tag{54}$$

This edge state has a positive group velocity v_F and thus corresponds to a "chiral right moving" edge state

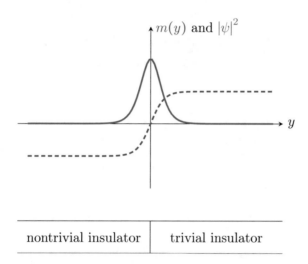

$m(y)$ and $|\psi|^2$

| nontrivial insulator | trivial insulator |

FIGURE 11. Schematic view of edge states at a Chern-trivial insulator interface. The mass $m(y)$ (blue dashed line) and the wavefunction amplitude $|\psi|^2$ (red continuous line) are drawn along the coordinate y orthogonal to the interface $y = 0$.

Similarly at the interface between a topological and a trivial insulator for the Kane–Mele model (37), (39) the function $d_1(k)$ changes sign at one of the λ_i TRIM points. Let us choose $m(y) = d_1[k = \lambda_0](y)$ with $m(y > 0) > 0$ and $m(y < 0) < 0$. The functions $d_{i \geq 2}$ are odd around λ_0 and up to a rotation of the local coordinates on the Brillouin zone we have $d_3(q) = q_x$ and $d_2(q) = -q_y$ and the linearized Hamiltonian around the TRIM λ_0 reads,

$$H_1(q) = q_x\,\Gamma_3 - q_y\,\Gamma_2 + m(y)\,\Gamma_1. \tag{55}$$

This Hamiltonian can be block-diagonalized in the basis (38) leading to the two surface solutions

$$\psi_{q_x,\uparrow}(x, y) \propto e^{-iq_x x} \exp\left[-\int_0^y m(y')\,dy'\right]\begin{pmatrix}0\\1\\0\\0\end{pmatrix} \tag{56a}$$

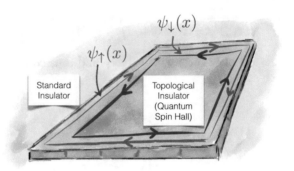

FIGURE 12. Illustration of edge states at the interface between a two-dimensional Kane–Mele insulator (Quantum Spin Hall Phase) and a standard insulator. A Kramers of counter-propagating states are present, represented here for simplicity by spin-up states (in red) and spin-down states (blue).

$$\psi_{q_x,\downarrow}(x,y) \propto \mathrm{e}^{+iq_x x} \exp\left[-\int_0^y m(y')\,\mathrm{d}y'\right] \begin{pmatrix} 0 \\ 0 \\ 0 \\ 1 \end{pmatrix}, \qquad (56b)$$

which consist of a Kramers pair of counter propagating states with opposite spins. A schematic representation of such a pair of edge states is represented in Figure 12. A mathematical discussion on this bulk-edge correspondence that goes far beyond the present introduction can be found in [31].

5.2. From topology in the bulk to critical semi-metals

The previous discussion also applies to the transition between bulk phases as a function of an external parameter such as chemical doping. We have already encountered this situation in the analysis of the Haldane model (32): on the phase diagram of Figure 6 the critical lines correspond to a situation where the gap h_z closes at one point of the Brillouin zone, and thus correspond to two-dimensional Dirac semi-metallic phases. One special point possesses two Dirac points: the case of graphene $M = \phi = 0$ where both parity and time-reversal symmetry are restored. A similar scenario holds at the transition from a topological insulator to an ordinary insulator both in two and three dimensions: as such a transition, the coefficient $d_1(\lambda_i)$ of the Kane–Mele model (37) vanishes. At the transition, the effective Hamiltonian describing the low electronic degree of freedom of the model is obtained by linearizing all three functions $d_i(\lambda_i + q)$ to first order in q. Up to dilatations and a rotation in momentum space the Hamiltonian reads:

$$H_{\text{Dirac}} = \Gamma_1 \, q_x + \Gamma_2 \, q_y + \Gamma_3 \, q_z, \tag{57}$$

where the Γ_i matrices satisfy a Clifford algebra: $\{\Gamma_i, \Gamma_i\} = 2\delta_{i,j}$: this is a massless three-dimensional Dirac equation. Hence the critical model at the transition between a topological and an ordinary insulator describes three-dimensional massless Dirac particles. This result is not longer valid when inversion symmetry is broken: the transition occurs then in two steps with a critical phase in between [43]. This critical region corresponds to a semi-metal where only two bands cross at two different points K, K'. Around each of these points, an effective 2×2 Hamiltonian is sufficient to describre the low energy electronic states, parametrized by Pauli matrices:

$$H_{\text{Weyl}} = \sigma_1 \, q_x + \sigma_2 \, q_y + \sigma_3 \, q_z. \tag{58}$$

Such a Hamiltonian does not describe excitations satisfying a Dirac equation: it is instead called a Weyl Hamiltonian, which has been proposed in particular as low energy Hamiltonian in correlated materials [44]. They possess remarkable surface states with unusual Fermi arcs.

Much of the recent activity in this direction have focused on the possibility to stabilize a three-dimensional Dirac semi-metal by crystalline symmetries to realize a three-dimensional analog of graphene. Indeed, a Dirac point described by a Hamiltonian (57) can be viewed as the superposition of two Weyl points such as equation (58) with opposite Chern numbers (see below). To prevent the annihilation of the two Weyl points, additional symmetries are required to forbid any hybridization between the associated bands (see [45] for a general discussion). Such a Dirac phase was recently proposed in Cd_3As_2 and tested experimentally [46].

5.3. Topological properties of semi-metallic phases

Much of the recent theoretical discussions of the stability and existence of semi-metallic phases have relied on a topological characterization of these phases, extending the previously discussed characterization of insulators. Indeed, we have already encountered that crossing between two bands appeared as analogs of Berry monopoles: in the analysis of the Chern index in Section 4.1.1, the point $h = 0$ corresponding to the two bands crossing acts as the source of the flux integrated on the surface spanned by $h(k)$, providing the Chern number. Similarly, the expression (16) of Berry curvature is found to be singular when two bands approach each other. Indeed, it is possible to attribute a topological number to such a band crossing points. This classification of topological properties of semi-metals proceeds along the same lines as the classification of defects in orderered media (vortices, dislocations, skyrmions, etc.) [2]: see in particular [47] for a related discussion. By definition in such a semi-metal a discrete set of points separate valence from conduction bands. If we exclude these points from the Brillouin zone, we obtain two well-separated sub-bundles of valence and conduction bands states, similarly to the insulating case. These vector bundles are now defined on the Brillouin zone punctured by these points. These two sub-bundles possess non-trivial topological properties as a consequence of the presence of these points: due to the modified topology of the punctured Brillouin zone, there now exist non-contractible loops ($d = 2$) or surfaces ($d = 3$) which encircle these points (see Figure 13). While the Berry curvature of the valence band might vanish everywhere in the punctured Brillouin zone, the Berry parallel transport of eigenstates along a path can be non trivial, associated with a winding independent of the path around the point.

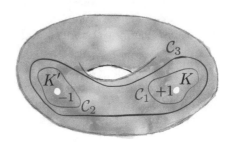

FIGURE 13. When the locations of the band crossing points are excluded from the Brillouin torus, new classes of non contractible loops (in d=2) exist, here represented by $\mathcal{C}_\infty, \mathcal{C}_\in$. They allow to define topological charges associated with these crossing points. The total charge over all crossing points necessarily vanishes: a loop such as $\mathcal{C}_\ni$ encircling all points is necessary contractible and can deformed to a point.

In two dimensions, we can define a winding number associated with any loop $\mathcal{C}$ encircling once the crossing point K, defined by

$$w_K = \frac{1}{\pi} \oint_{\mathcal{C}} A, \tag{59}$$

where A is the Berry connection form of the valence band. Note that necessarily, the sum of winding numbers of the different crossing points in the Brillouin zone must vanish: it corresponds to the winding number along a path encircling all points, which is necessarily contractible (see Figure 13). In the case of graphene, the two Dirac points are associated with opposite winding numbers ± 1. These numbers are topological in the following sense: they are robust with respect to any perturbation that doesn't open a gap at a Dirac point. From the Haldane model (32) we can identify these perturbations as those preserving time-reversal and parity symmetry. Within the ensemble of such perturbations, the topological numbers associated with each point is robust. The only way to modify the topology of the band structure is through the merging or fusion of these points. Such topological transitions have been studied in the context of graphene in [48] and observed in cold atoms lattices [49]. Similar reasoning allows to define the topological number characterizing a semi-metallic point K in three dimensions as the Chern number

$$n_K = \frac{1}{2\pi} \oint_S F, \tag{60}$$

where S is now a surface enclosing the point K, and the Berry curvature is defined on the valence bands. For a model with two bands crossing, described by a general Weyl Hamiltonian $H_{\text{Weyl}}(k = K + q) = v_{ij}q_i\sigma_j$, the calculation goes exactly along the lines of the analysis of Section 4.1.1, except that the integral over momentum q does run over the two-dimensional sphere encircling K instead of the two-dimensional Brillouin torus. We obtain a Chern number given by [45] $n(K) = \text{sign}\,(\det\,[v_{ij}])$.

6. Conclusion and perspectives

At the point of concluding these notes, it is worth mentioning a few subjects that have not been covered here by lack of time or knowledge. The ideas sketched in these notes have led to a classification of all 10 topological properties of gapped phases, whether insulating or superconducting, within a single particle framework. Such a classification can be obtained either by studying the robustness of surface states with respect to disorder, or by studying directly the topological classes of vector bundles within K-theory. Of great natural interest is the incorporation of interactions within this picture. Moreover, in an attempt to describe the origin of topological properties of bands, we have neglected the description of the essentials properties of the associated Dirac surface states and thus the physical properties of these materials. This constitute a vast subject, and the interested reader should refer to the serious reviews already mentioned [11, 12, 13, 14].

Appendix: Two useful trivializations of the Bloch bundle

A.1. Trivialization by Fourier transform

An example of trivialization of the Bloch bundle $\mathcal{H}$ is provided by starting from a choice of unit cell $\mathcal{F}$ of the crystal $\mathcal{C}$ (see Section 2). Such a unit cell is constituted of N points $x_\alpha, \alpha = 1, \ldots, N$. Let us denote by $|e_\alpha^{\text{I}}(k)\rangle$ the Fourier transforms of functions δ_{x,x_α} concentrated at these points x_α. These Fourier transforms constitute a smooth set of sections of the Bloch bundle, which furthermore constitute a basis of each fiber $\mathcal{H}_k$: Bloch states decompose uniquely as $|\varphi(k)\rangle = \sum_\alpha \varphi(k; x_\alpha)|e_\alpha^{\text{I}}(k)\rangle$. We shall denote by ∇^{I} the flat connection associated to the trivialization $k \mapsto e_k^{\text{I}i}$ of $\mathcal{H}$. This trivialization of $\mathcal{H}$ and its associated connection depends on the associated choice of unit cell $\mathcal{F}$. Another choice $\mathcal{F}'$ of unit cell is related to $\mathcal{F}$ up to a possible relabeling of points by $x'_\alpha = x_\alpha + \gamma_\alpha$ with γ_α a vector of the Bravais lattice $\gamma_\alpha \in \Gamma$. Both trivialization are then related by the simple proportionality relation $|e_\alpha^{\prime\text{I}}(k)\rangle = e^{ik\cdot\gamma_\alpha}|e_\alpha^{\text{I}}(k)\rangle$, and the associated flat connections only differ (as expected for two flat connections) by a differential 1-form:

$$\nabla'^{\mathrm{I}} = \nabla^{\mathrm{I}} - \mathrm{i} \sum_\alpha |e^{\mathrm{I}}_\alpha(k)\rangle\langle e^{\mathrm{I}}_\alpha(k)| \, dk \cdot \gamma_\alpha. \tag{61}$$

A.2. Trivialization from periodic functions

A second choice of trivialization appears more canonical as it depends in a trivial way of the arbitrariness of this definition (the choice of origin of space). It corresponds to the common writing of Bloch functions $|\varphi(k)\rangle \in \mathcal{H}_k$ as a functions of periodic functions

$$\varphi(k; x) = \mathrm{e}^{-\mathrm{i}k\cdot(x-x_0)} u(k; x), \tag{62}$$

where x is any point in the crystal $\mathcal{C}$ and x_0 an arbitrary origin of the Euclidean space, not necessarily in the crystal. In this expression, $u(k; .)$ is a periodic function on the Bravais lattice: $u(k; x + \gamma) = u(k; x)$ for $\gamma \in \Gamma$. Hence equation (62) establishes a relation between each fiber $\mathcal{H}_k$ of quasi-periodic Bloch functions and the vector space $\ell^2(\mathcal{C}/\Gamma)$ of Γ-periodic function on the crystal $\mathcal{C}$. Such a function $\varphi(k; x)$ does not possess the periodicity of the reciprocal lattice as $u(k + G; x) = \mathrm{e}^{\mathrm{i}G\cdot(x-x_0)} u(k; x)$ for $G \in \Gamma^\star$. Hence the writing (62) establishes an identification between the Bloch bundle $\mathcal{H}$ and the quotient of the trivial bundle $\mathbb{R}^d \times \ell^2(\mathcal{C}/\Gamma)$ of periodic function u indexed by a real vector $k \in \mathbb{R}^d$, by the action of the reciprocal lattice $\Gamma^\star$ (which preserves the scalar produce on $\ell^2(\mathcal{C}/\Gamma)$):

$$(k, u(x)) \longmapsto (k + G, \, \mathrm{e}^{\mathrm{i}G\cdot(x-x_0)} u(x)), \text{ for } G \in \Gamma^\star. \tag{63}$$

The bundle $\mathbb{R}^d \times \ell^2(\mathcal{C}/\Gamma)$ of periodic function u indexed by a real vector is trivial in the sense defined previously: it possesses a natural flat connection obtained by choosing a basis of periodic functions $\ell^2(\mathcal{C}/\Gamma)$ independent of k. A natural choice for this basis is $u^\alpha(x) = \sum_{\gamma \in \Gamma} \delta_{x, x_\alpha + \gamma}$, $\alpha = 1, \ldots, N$ where the x_α are the points of a unit cell $\mathcal{F}$. It should be noted that the u^α are independent of this choice of $\mathcal{F}$: they are indexed by sub-lattices, and not by a choice of point x_α in each sub-lattice. This basis is pulled back as a basis of Bloch functions in each $\mathcal{H}_k$ as $e^{\mathrm{II}}_\alpha(k; x) = \mathrm{e}^{-\mathrm{i}k\cdot(x-x_0)} u^i(x)$ The connection ∇^{II} is now defined by the exterior derivative of its sections in this basis (see equation(18)).

Acknowledgment

I thank Pierre Delplace, Michel Fruchart, Thibaut Louvet and above all Krzysztof Gawedzki, from whom I've learned through friendly and stimulating discussions most of what is included in this review.

References

[1] Dirac, P.A.M.: Quantised singularities in the electromagnetic field. Proceedings of the Royal Society of London. Series A, Containing Papers of a Mathematical and Physical Character **133**, 60–72 (1931).

[2] Toulouse, G., and Kléman, M.: Principles of a classification of defects in ordered media. J. Physique Lett. **37**, L–149 (1976).

[3] Thouless, D.J., Kohmoto, M., Nightingale, M.P., and den Nijs, M.: Quantized Hall conductance in a two-dimensional periodic potential. Phys. Rev. Lett. **49**, 405–408 (1982).

[4] Avron, J.E., and Seiler, R.: Quantization of the Hall conductance for general, multiparticle Schrödinger Hamiltonians. Phys. Rev. Lett. **54**, 259 (1985).

[5] Niu, Q., Thouless, D.J., and Wu, Y.-S.: Quantized hall conductance as a topological invariant. Phys. Rev. B **31**, 3372 (1985).

[6] Wen, X.-G.: Topological orders in rigid states. Int. J. Mod. Phys. B **4**, 239 (1990).

[7] Haldane, F.D.M.: Model for a quantum Hall effect without landau levels: Condensed-matter realization of the "parity anomaly". Phys. Rev. Lett. **61**, 2015–2018 (1988).

[8] Berry, M.V.: Quantal phase factors accompanying adiabatic changes. Proceedings of the Royal Society of London. A. Mathematical and Physical Sciences **392**, 45–57 (1984).

[9] Simon, B.: Holonomy, the quantum adiabatic theorem, and Berry's phase. Phys. Rev. Lett. **51** 2167–2170 (1983).

[10] Kane, C.L., and Mele, E.J.: Z_2 topological order and the quantum spin Hall effect. Phys. Rev. Lett. **95**, 146802 (2005).

[11] Hasan, M.Z., and Kane, C.L.: Colloquium: Topological insulators. Rev. Mod. Phys. **82**, 3045–3067 (2010).

[12] Qi, X.-L., and Zhang, S.-C.: Topological insulators and superconductors. Rev. Mod. Phys. **83**, 1057 (2011).

[13] Bernevig, B.A., and Hughes, T.L.: *Topological Insulators and Topological Superconductors*. Princeton University Press, 4, 2013.

[14] Franz, M., and Molenkamp, L., editors: *Topological Insulators*. Elsevier, 2013.

[15] Blount, E.I.: *Formalisms of band theory*. In F. Seitz and D. Turnbull, editors, *Solid State Physics*, volume 13, pages 305–373. Academic Press, 1962.

[16] Aharonov, Y., and Bohm, D.: Significance of electromagnetic potentials in quantum theory. Phys. Rev. **115**, 485 (1959).

[17] Webb, R.A., Washburn, S., Umbach, C.P., and Laibowitz, R.P.: Observation of h/e Aharonov–Bohm oscillations in normal-metal rings. Phys. Rev. Lett. **54**, 2696 (1985).

[18] Kato, T.: On the adiabatic theorem of quantum mechanics. J. Phys. Soc. Jpn **5**, 435 (1950).

[19] Aharonov, Y., and Anandan, J.: Phase change during a cyclic quantum evolution. Phys. Rev. Lett. **58**, 1593 (1987).

[20] Anandan, J., and Aharonov, Y.: Geometry of quantum evolution. Phys. Rev. Lett. **65**, 1697 (1990).

[21] Nakahara, M.: *Geometry, Topology and Physics*. Taylor & Francis, 2nd edition, 2003.

[22] Xiao, D., Chang, M.-C., and Niu, Q.: Berry phase effects on electronic properties. Rev. Mod. Phys. **82**, 1959–2007 (2010).

[23] Fruchart, M., Carpentier, D., and Gawedzki, K.: Parallel transport and band theory in crystals. Eur. Phys. Lett. **106**, 60002 (2014).

[24] Bena, C., and Montambaux, G.: Remarks on the tight-binding model of graphene. New Journal of Physics **11**, 095003 (2009).

[25] Fruchart, M., and Carpentier, D.: An introduction to topological insulators. Comptes Rendus Physique **14**, 779 (2013).

[26] Piéchon, F., and Suzumura, Y.: Inversion symmetry and wave-function-nodal-lines of dirac electrons in organic conductor α-(BEDT-TTF)$_2$I$_3$. arXiv:1309.3495.

[27] Fuchs, J.N., Piéchon, F., Goerbig, M.O., and Montambaux, G.: Topological Berry phase and semiclassical quantization of cyclotron orbits for two-dimensional electrons in coupled band models. Eur. Phys. J. B **77**, 351–362 (2010).

[28] Sticlet, D., Piéchon, F., Fuchs, J.-N., Kalugin, P., and Simon, P.: Geometrical engineering of a two-band chern insulator in two dimensions with arbitrary topological index. Phys. Rev. B **85**, 165456 (2012).

[29] Sakurai, J.J.: *Modern Quantum Mechanics*. Addison Wesley, 1 edition, 9 1993.

[30] Fu, L., and Kane, C.L.: Topological insulators with inversion symmetry. Phys. Rev. B **76**, 045302 (2007).

[31] Graf, G.M., and Porta, M.: Bulk-edge correspondence for two-dimensional topological insulators. Comm. Math. Phys. **324**, 851 (2013).

[32] Fu, L., Kane, C.L., and Mele, E.J.: Topological insulators in three dimensions. Phys. Rev. Lett. **98**, 106803 (2007).

[33] Fu, L., and Kane, C.L.: Time reversal polarization and a Z_2 adiabatic spin pump. Phys. Rev. B **74**, 195312 (2006).

[34] Roy, R.: Z_2 classification of quantum spin Hall systems: An approach using time-reversal invariance. Phys. Rev. B **79**, 195321 (2009).

[35] Moore, J.E., and Balents, L.: Topological invariants of time-reversal-invariant band structures. Phys. Rev. B **75**, 121306 (2007).

[36] Soluyanov, A.A., and Vanderbilt, D.: Wannier representation of Z_2 topological insulators. Phys. Rev. B **83**, 035108 (2011).

[37] Qi, X.-L., Hughes, T.L., and Zhang, S.-C.: Topological field theory of time-reversal invariant insulators. Phys. Rev. B **78**, 195424 (2008).

[38] Wang, Z., Qi, X.-L., and Zhang, S.-C.: Equivalent topological invariants of topological insulators. New Journal of Physics **12** (6), 065007 (2010).

[39] Bernevig, B.A., Hughes, T.L., and Zhang, S.-C.: Quantum Spin Hall Effect and Topological Phase Transition in HgTe Quantum Wells. Science **314**, 1757–1761 (2006).

[40] Liu, C., Hughes, T.L., Qi, X.L., Wang, K., and Zhang, S.C.: Quantum spin hall effect in inverted type-ii semiconductors. Phys. Rev. Lett. **100**, 236601 (2008).

[41] Bastard, G.: *Wave Mechanics applied to Semiconductor Heterostructures*. Les Editions de Physique, 1990.

[42] Jackiw, R., and Rebbi, C.: Solitons with fermion number 1/2. Phys. Rev. D **13** (13), 3398–3409 (1976).

[43] Murakami, S.: Phase transition between the quantum spin hall and insulator phases in 3d: emergence of a topological gapless phase. New Journal of Physics **9**, 356 (2007).

[44] Turner, A.M., and Vishwanath, A.: *Beyond Band Insulators: Topology of Semimetals and Interacting Phases.* Chap. 11 in [14].

[45] Young, S.M., Zaheer, S., Teo, J.C.Y., Kane, C.L., Mele, E.J., and Rappe, A.M.: Dirac semimetal in three dimensions. Phys. Rev. Lett. **108**, 140405 (2012).

[46] Wang, Z., Weng, H., Wu, Q., Dai, X., and Fang, Z.: Three-dimensional dirac semimetal and quantum transport in cd3as2. Phys. Rev. B **88**, 125427 (2013).

[47] Volovik, G.E.: *The Universe in a Helium Droplet.* Oxford University Press, 2003.

[48] Montambaux, G., Piéchon, F., Fuchs, J.-N., and Goerbig, M.O.: Merging of dirac points in a 2d crystal. Phys. Rev. B **80**, 153412 (2009).

[49] Tarruell, L., Greif, D., Uehlinger, T., Jotzu, G., and Esslinger, T.: Creating, moving and merging dirac points with a fermi gas in a tunable honeycomb lattice. Nature **483**, 302 (2012).

David Carpentier
Laboratoire de Physique
Ecole Normale Supérieure de Lyon
46, Allée d'Italie
F-69007 Lyon, France
e-mail: `David.Carpentier@ens-lyon.fr`

Printed in the United States
By Bookmasters